Clemens Chan-Braun

Turbulent open channel flow, sediment erosion and sediment transport

Dissertationsreihe am Institut für Hydromechanik
Karlsruher Institut für Technologie (KIT)

Heft 2012/1

Turbulent open channel flow, sediment erosion and sediment transport

by
Clemens Chan-Braun

Dissertation, genehmigt von der Fakultät für Bauingenieur-, Geo- und Umweltwissenschaften des Karlsruher Instituts für Technologie (KIT), 2012
Referenten: Prof. Dr. Markus Uhlmann, Prof. Dr. habil. Michael Manhart, Prof. Dr.-Ing. Bettina Frohnapfel, Prof. Dr.-Ing. Manuel García-Villalba, Prof. i.R. Dr. habil. Wolfgang Rodi, Prof. Dr.-Ing. Karl Schweizerhof

Impressum

Karlsruher Institut für Technologie (KIT)
KIT Scientific Publishing
Straße am Forum 2
D-76131 Karlsruhe
www.ksp.kit.edu

KIT – Universität des Landes Baden-Württemberg und nationales Forschungszentrum in der Helmholtz-Gemeinschaft

KIT Scientific Publishing 2012
Print on Demand

ISSN 1439-4111
ISBN 978-3-86644-900-8

Turbulent open channel flow, sediment erosion and sediment transport

Zur Erlangung des akademischen Grades eines

DOKTOR-INGENIEURS

von der Fakultät für

Bauingenieur-, Geo- und Umweltwissenschaften

der Universität Fridericiana zu Karlsruhe (TH)

genehmigte

DISSERTATION

von

Clemens Chan-Braun

geboren in Bad Dürkheim in der Pfalz

Tag der mündlichen Prüfung: 10. Februar 2012

Hauptreferent: Prof. Dr. Markus Uhlmann
Korreferent: Prof. Dr. habil. Michael Manhart
Prof. Dr.-Ing. Bettina Frohnapfel
Prof. Dr.-Ing. Manuel García-Villalba
Prof. i.R. Dr. habil. Wolfgang Rodi
Prof. Dr.-Ing. Karl Schweizerhof

Karlsruhe (2012)

To Linda

Abstract

This thesis is a contribution towards a deeper understanding of turbulent open channel flow, sediment erosion and sediment transport. The thesis provides an analysis of high-fidelity data from three different flow configurations: (i) open channel flow over an array of fixed spheres, (ii) open channel flow with mobile eroding spheres, (iii) open channel flow with sediment transport of many mobile spheres. The data is generated by direct numerical simulation using an immersed boundary method to resolve the surface of fixed and mobile particles. The simulations provide a detailed picture of the flow field and particle related quantities. The results show, that simplified considerations of the relation of flow structures to force and torque on particles are useful to explain characteristics of the force and torque on particles. For the present flow configurations, the time and velocity scales of force and torque are of the order of outer flow units and the correlation of flow structures to force and torque fluctuations are of the order of the channel height. The onset of sediment erosion is discussed by the results of simulations with fixed and with mobile eroding spheres. Conditionally averaged flow fields and instantaneous flow structures related to the onset of sediment erosion agree with experimental findings from the literature. However, the prediction of the onset of sediment erosion by a critical Shields number based on instantaneous lift on a fixed sphere are not in line with the experimental evidence in the literature. An explanation of this discrepancy could be that this approach neglects possible collective effects. The few simulations with mobile particles indicate that such collective effects may play an important role at the given parameter range. Simulations of sediment transport reveal a strong influence of the mobile particles on the flow field statistics. It is found that the Rouse formula provides a good approximation of the obtained density profiles of the present cases, when the exponent of the original definition is reduced by 20% to 35%.

Zusammenfassung

Die vorliegende Arbeit beschäftigt sich mit Fragestellungen zur turbulenten Strömung in offenen Gerinnen, zum Erosionsbeginn von Sedimentpartikel und zum Sedimenttransport. Die Fragestellungen sind von besonderer Bedeutung für den Fluss- und Wasserbau. Zum Beispiel werden für die Kontrolle von Flüssen und den sich darin befindenden hydraulischen Bauwerken Gesetzmäßigkeiten benötigt, die das Strömungsverhalten, den Beginn der Sedimenterosion sowie den Sedimenttransport beschreiben. Die meisten der derzeit verwendeten Methoden basieren auf empirischen Annahmen und zeitlich sowie räumlich gemittelten Strömungsgrößen. Die Genauigkeit der Ansätze ist jedoch unbefriedigend. Dies ist zum Teil in einem unzureichenden Verständnis der zugrunde liegenden Prozesse begründet, was wiederum an einem Fehlen an Untersuchungen der Prozesse mit einer ausreichenden Auflösung in Zeit und Raum liegt. Die vorliegenden Arbeit versucht diese Lücken teilweise zu schließen, indem hoch genaue Daten von verschiedenen direkten numerischen Simulationen präsentiert und diskutiert werden. Eine effiziente Diskretisierung der Partikeloberfläche wird hierbei mit Hilfe einer Methode realisiert, die die Definition von Randbedingung an einem beliebigen Punkt im Berechnungsgebiet ermöglicht. Drei verschiedene Konfigurationen werden untersucht: (i) die offene Gerinneströmung über unbewegliche Kugeln, (ii) die offene Gerinneströmung mit beweglichen erodierenden Kugeln, (iii) die offene Gerinneströmungen mit Transport von vielen beweglichen Kugeln. Die Analyse der Simulationen konzentriert sich auf drei Aspekte: (*a*) die Charakterisierung von Kraft und Moment auf Partikel und den damit verbundenen Strömungsstrukturen, (*b*) die Charakterisierung der Vorgänge die zum Beginn der Erosion von Partikeln führen, basierend auf den Ergebnissen mit unbeweglichen und mit beweglichen Kugeln, (*c*) den Einfluss von sich bewegenden Kugeln auf die Statistiken des Strömungsfelds und die der Partikel. Im Folgenden werden die Ergebnisse kurz zusammengefasst.

Offene Gerinneströmungen über unbewegliche Kugeln Zwei direkte numerische Simulationen einer offenen Gerinneströmung über unbewegliche Kugeln in quadratische Anordnung wurden durchgeführt. Die Reynoldszahl, basierend auf der effektiven Kanalhöhe und der mittleren Durchschnittsgeschwindigkeit, beträgt in beiden Simulationen 2900. Die Konfigurationen unterscheiden sich in Bezug auf die Größe der Kugeln und des hydraulischen Regimes der turbulenten Strömung. In der einen Konfiguration sind die Kugeln klein, der Durchmesser der Kugeln ist 10,7 viskose Längeneinheiten, und die Strömung befindet sich in der Nähe des hydraulisch glatten Strömungsregimes. In der anderen Konfiguration sind die Kugel mehr als dreimal größer (49,3 viskose Längeneinheiten) und die Strömung befindet sich im Übergangsbereich zwischen hydraulisch glatt und hydraulisch rau. Die Ergebnisse der Simulationen werden zunächst in Hinblick auf das Strömungsfeld untersucht und mit Ergebnissen aus der Literatur verglichen. Dazu werden die Statistiken des Strömungsfelds, der Einfluss der Rauigkeit, sowie das zeitlich gemittelte dreidimensionale Strömungsfeld besprochen. Der eigentliche Schwerpunkt der Analyse ist jedoch die Charakterisierung der Kraft und des Moments auf die Kugeln, sowie die Charakterisierung der Strömungsstrukturen die Kraft und Momentfluktuationen bewirken. Analysiert werden insbesondere die Statistiken von Kraft und Moment auf die Partikel, die Korrelation von Kraft und Moment in der Zeit und in der Raum-Zeit, sowie die Korrelationen zwischen Kraft- und Momentfluktuationen und Strömungsfluktuationen. Die Statistiken der Partikelkräfte ergeben positive Mittelwerte der Kraftkomponente in Strömungsrichtung (Widerstandskraft) sowie der Kraftkomponente normal zur Wand (Auftriebskraft), der Mittelwert des Moments in Spannweitenrichtung ist negativ. Diese Mittelwerte entstehen zu einem Großteil im oberen Bereich der Kugeln. Die Intensität der Kraftfluktuationen nimmt mit der Größe der Partikel deutlich zu. Im Gegensatz dazu zeigen die einzelnen Komponenten der Momentfluktuationen unterschiedliche Tendenzen. Um die Strömungsstrukturen zu untersuchen, die mit den Kraft- und Momentfluktuationen in Verbindung stehen, werden die Korrelationen zwischen diesen Größen untersucht. Die Korrelationen haben in beiden Simulationen Längenskalen von der Größenordnung der Gerinnehöhe. Ebenso sind die Zeitskalen der zeitlichen Korrelationen von Kraft und Moment vergleichbar zu den Zeitskalen der Außensströmung. Die Ergebnisse der Raum–Zeit Korrelation ergeben, dass sich die Konvektionsgeschwindigkeit der Kraft und des Moments mit 46%-71% der mittleren Strömungsgeschwindigkeit des Fluids fortbewegen. Im Fall der großen Kugeln nehmen die Konvektionsgeschwindigkeiten ab, was sich mit einem größeren Einfluss der Rauigkeit begründen lassen könnte.

Beginn der Sedimenterosion Diese Arbeit versucht die Ereignisse zu charakterisieren, die zum Beginn von Sedimenterosion führen. Dies geschieht auf Grundlage der Simulationen über unbewegliche Kugeln sowie der Simulationen mit beweglichen erodierenden Kugeln. Zuerst werden verschiedene Methoden untersucht, um den Beginn von Sedimenterosion mit Hilfe von Kraft und Moment auf ein unbewegtes Partikel zu prognostizieren. Es zeigt sich, dass Prognosen basierend auf der Widerstandskraft, der Auftriebskraft oder auf dem Moment in Spannweitenrichtung zu unterschiedlichen Ergebnissen führen können. Methoden die auf abgeleitete Größen zurückgreifen, z.B. auf die Kraft tangential zur Ebene durch die Auflager des Partikels, oder auf das Moment um die Achse durch die Auflager, sind hier äquivalent zu Methoden die auf der Widerstandskraft basieren. Die Analyse ergibt allerdings keinen Schwellenwert der besonders für die Definition des Beginns der Sedimenterosion geeignet scheint. Danach werden die Ereignisse untersucht, die zu sehr hohen Widerstandskräften, Auftriebskräften oder sehr niedrigen Momenten in Spannweitenrichtung führen. Konditioniert gemittelte Zeitsignale von Widerstandskraft, Auftriebskraft und Moment in Spannweitenrichtung ergeben stark unterschiedliche Ergebnisse in Abhängigkeit von der Konditionierungsvariablen. Es fällt auf, dass der Charakter der konditioniert gemittelten Zeitsignale große Ähnlichkeiten zu den entsprechenden Korrelationen zwischen den Zeitsignalen der gemittelten Größe und der Größe der Konditionierung aufweist. Instantane Strömungsfelder um einen Partikel der eine sehr hohe Widerstandskraft erfährt, zeigen oft eine Struktur mit hohen und eine mit niedrigen Geschwindikeitsfluktuationen in Strömungsrichtung. Ebenso befinden sich oft Strukturen mit hoch negativem Druck in der Nähe des Partikels. Die Region um den Partikel ist durch eine hohe turbulente Aktivität geprägt. Um eine quantitative Analyse der Ereignisse zu ermöglichen, werden die auf eine hohe Widerstandskraft konditionierten Strömungsfelder gemittelt. Die konditioniert gemittelten Geschwindigkeitsfelder in Strömungsrichtung zeigen direkt über dem Partikel mit einer hohen Widerstandskraft eine Region mit positiver Geschwindigkeitsfluktuation, die sich in Strömungsrichtung über mehrere Kanalhöhen erstreckt. Strömungsfluktuationen in Richtung der Wand dominieren die mittlere vertikale Strömungskomponente um den Partikel. Im Mittel strömt das Fluid in Spannweitenrichtung vom Partikel weg. Einmal mehr findet man, dass die Form der konditioniert gemittelten Strömungsfelder große Ähnlichkeiten zu den Korrelationen zwischen Partikelkraft/-moment und Strömungsfeld aufweist.

Die Wahrscheinlichkeit des Erosionsbeginns eines unbeweglichen Partikels kann mit Hilfe einer Shieldszahl abgeschätzt werden, die durch die instantane Auftriebskraft definiert ist. Diese Abschätzung ergibt kritische Shieldszahlen von kleinerem Wert für die größeren Partikel was im Gegensatz zu den experimentellen Erfahrungen steht. Ein Grund für diesen Unterschied könnte die Vernachlässigung von kollektiven Effekten bei dieser Vorgehensweise sein. Die wenigen durchgeführten Simulationen mit bewegten Partikeln deuten an, dass solche Effekte eine große Rolle spielen könnten. Die Simulationen zeigen auch, dass Partikel während der Erosion stark in Wandnormalenrichtung beschleunigt werden und ebenso entlang der Achse in Spannweitenrichtung. In Übereinstimmung mit experimentellen Ergebnissen werden hierbei vertikale Geschwindigkeiten von der Größenordnung der Schubspannungsgeschwindigkeit erreicht.

Sedimenttransport in offenen Gerinneströmungen Direkte numerische Simulationen der offenen Gerinneströmung mit Sedimenttransport von mehreren beweglichen Kugeln wurden durchgeführt. Die Strömungskonfigurationen sind hierbei ähnlich zu der zuvor untersuchten Gerinneströmung über kleine unbewegliche Kugeln. Der Unterschied ist, dass nun 2000 bzw. 9126 zusätzliche bewegliche Kugeln in das Strömungsfeld induziert wurden. Dies entspricht einem globalen Volumenverhältnis von $3{,}0 \cdot 10^{-3}$ und $1{,}4 \cdot 10^{-2}$. Gravitation ($|g|\, h/U_{bh}^2 = 0{,}7$) wird in Richtung der Wand berücksichtigt und das Verhältnis zwischen Partikeldichte und Fluiddichte ist 1,7.

Aufgrund der Gravitation akkumulieren die Partikel in der Nähe der Wand. In dem gewählten Parameterraum führt jedoch der turbulente Strömungseinfluss zu einem Zyklus von Partikelsuspension und Partikelablagerung. Dieser resultiert in einem mit der Entfernung zur Wand abnehmenden Konzentrationsprofil der Partikel. Die Ergebnisse zeigen, dass die Präsenz der sich bewegenden Partikel einen großen Einfluss auf das mittlere Strömungsprofil und die Profile der turbulenten Schwankungen des Fluids hat. Die mittlere Geschwindigkeit der Partikel ist im Mittel kleiner als die des Fluids. Beide Erkenntnisse bestätigen vorhergehende Experimente und legen den Schluss nahe, dass die Trägheit der Partikel, die Auswirkung der finiten Größe der Partikel und der Reynoldszahl, sowie die Gravitation eine große Bedeutung für die untersuchte Strömungskonfiguration hat. Einige der Mechanismen der Interaktion zwischen Partikel und turbulenter Strömung werden diskutiert.

Acknowledgements

This work was supported by the German Research Foundation (DFG) under project JI 18/19-1. The computations have been carried out at the Steinbuch Centre for Computing (SCC) of Karlsruhe Institute of Technology and at the Leibniz Supercomputing Centre (LRZ) of the Bavarian Academy of Sciences and Humanities. Within this PhD project scholarships to attain conferences and to arrange a scientific exchange visit abroad were granted by the German Academic Exchange Service (DAAD), by the European Commission (Marie-Curie) and by the Karlsruhe House of Young Scientists (KHYS). The support from these institutions is gratefully acknowledged.

I am much obliged to Prof. Markus Uhlmann for his excellent supervision and scientific guidance of my work. Without his involvement this thesis could not have been accomplished in the present form. I would like to thank Prof. Michael Manhart for his friendly disposition as the second referee of this thesis and Prof. Bettina Frohnapfel and Prof. Karl Schweizerhof for their disposition as members of my PhD committee. I am grateful to Prof. Wolfgang Rodi for his encouragement from the very beginning and for the possibility to work in his group. I am also thankful to late Prof. Gerhard Jirka for his interest in the project. Without his enthusiasm I would have not been able to carry out the research. I am particularly indebted to Prof. Manuel García-Villalba, not only for sharing his understanding of turbulence and science in many discussions but also for his continuous support and advice in difficult situations.

I am grateful to the people I was able to meet during my time at the Institute for Hydromechanics, in particular to Aman, Fred, Todor, Antje, Conny, Herlina, Martin, Michael, Tobias, Wernher, Tim, Hannes, Thorsten and Jan.

The hours I spent with the members of the 08/15–Jazzband in (dis)harmony have been highly beneficial to balance the work load. As were the frequent sessions of physical exercise with Carsten Hasberg. To all of you many thanks for your friendship.

Last but not least, I would like to thank both sides of my family for their encouragement during the past years. Vielen Dank an meine Familie für ihre Unterstützung in den letzten Jahren auch über weite Distanzen hinweg. Special thanks to Linda for her enduring love and backup throughout the course of this thesis.

Related publications

The content of this theses is based in parts on the following references, published in the course of this PhD thesis. The contributions of the co-authors is highly acknowledged.

Braun, C., García-Villalba, M., Jirka, G., and Rodi, W. (2008). Impact of turbulent flow on large spherical roughness elements. In *EUROMECH Fluid Mech. Conf. 7*, University of Manchester, United Kingdom.

Braun, C. (2009). *First results on the impact of turbulent flow on fixed large spherical roughness elements.* Report 840, Institute for Hydromechanics, University of Karlsruhe (TH), Germany.

Braun, C., García-Villalba, M., and Uhlmann, M. (2009). A computational study of the hydrodynamics forces on a rough wall. In Eckhardt, B., editor, *Advances in Turbulence XII*, Springer Proc. Physics 132, p. 929.

Braun, C., García-Villalba, M., and Uhlmann, M. (2009). Particle force generation in a turbulent open channel flow. In *Proc. 33rd IAHR Cong.: Water engineering for a sustainable environment*, pp. 44–50, Vancouver, Canada.

Chan-Braun, C., García-Villalba, M., and Uhlmann, M. (2010). Direct numerical simulation of sediment transport in turbulent open channel flow. In Nagel, W., Kröner, D., and Resch, M., editors, *High performance computing in science and engineering '10*. Springer.

Chan-Braun, C., García-Villalba, M., and Uhlmann, M. (2010). Numerical simulation of fully resolved particles in rough-wall turbulent open channel flow. In Balachandar, S. and Curtis, J. S., editors, *Proc. 7th Int. Conf. Multiphase Flow*, ICMF 2010, Tampa, USA.

Chan-Braun, C., García-Villalba, M., and Uhlmann, M. (2010). Numerical simulation of the onset of sediment erosion. In Adams, N. A., Maier, I., Pernpeintner, A., Su, L., and Wall, W., editors, *EUROMECH Fluid Mech. Conf. 8*, Bad Reichenhall, Germany.

Chan-Braun, C., Strehle, H., García-Villalba, M., and Uhlmann, M. (2010). Direct numerical simulation of sediment erosion in an open channel flow. In Balachandar, S. and Curtis, J. S., editors, *Gallery of Multiphase Flow, 7th Int. Conf. Multiphase Flow*, ICMF 2010, Tampa, USA.

Chan-Braun, C., García-Villalba, M., and Uhlmann, M. (2011). Direct numerical simulation of rough wall open channel flow. In *Proc. 7th Int. Symp. Turbul. Shear Flow Phen.*, Ottawa, Canada.

Chan-Braun, C., García-Villalba, M., and Uhlmann, M. (2011). Force and torque acting on particles in a transitionally rough open channel flow. *J. Fluid Mech.*, 684:441–474.

Sometimes you have to play a long time
to be able to play like yourself.

Miles Davis

Contents

Nomenclature

Mathematical symbols

$\langle \cdot \rangle$	averaging in time and in spanwise and streamwise direction
$\langle \cdot \rangle_{\text{cond}}$	conditional averaging as defined in (4.5)
$\langle \cdot \rangle_{t,x,y,z,b,p}$	according to sub-index: averaging over time (t), along a certain coordinate direction (x, y, z), with respect the periodically repeating boxes of geometry (b) or/and over the particle related data (p), respectively
$\cdot$	dot product of two vectors
$\times$	cross product of two vectors
∇	Nabla operator, i.e. in Cartesian coordinates, $\nabla = (\partial/\partial x, \partial/\partial y, \partial/\partial z)$
∇^2	Laplace operator, i.e. in Cartesian coordinates, $\nabla^2 = (\partial^2/\partial x^2, \partial^2/\partial y^2, \partial^2/\partial z^2)$
ϕ^+	a quantity, here ϕ, normalised with wall units, i.e. involving u_τ and/or ν
$\widehat{\phi}$	Fourier transformed quantity ϕ
$\widehat{\phi}^*$	conjugate complex of $\widehat{\phi}$
ϕ'	fluctuation of a quantity ϕ, e.g. $\phi' = \phi - \langle \phi \rangle$
ϕ''	fluctuation of time and box averaged quantity ϕ to its time and plane-averaged value, i.e. $\phi'' = \langle \phi \rangle_{tb} - \langle \phi \rangle$
ϕ_{max}	maximal value of a quantity, i.e. $\phi_{\text{max}} = \max(\phi)$, or a quantity related to a maximum, i.e. $\phi(\tau_{\text{max}}) = \max(\phi)$
ϕ_{min}	minimal value of a quantity, i.e. $\phi_{\text{min}} = \min(\phi)$, or a quantity related to a minimum, i.e. $\phi(\tau_{\text{min}}) = \min(\phi)$
ϕ_{rms}	root mean square value of a quantity ϕ
ϕ_{thres}	threshold related to a quantity ϕ

Greek symbols

α	angle to horizontal axis

α_{bin}	stretching factor (cf. equation D.1)
β	parameter in Rouse formula (cf. equation 5.1)
σ_ϕ	standard deviation of a quantity ϕ
δ_ν	viscous length, $\delta_\nu = \nu/u_\tau$
ΔU	velocity offset due to roughness in law of the wall
Δ_x	grid spacing in streamwise direction
Δ_y	grid spacing in wall-normal direction
Δ_z	grid spacing in spanwise direction
Γ	control surface bounding a certain volume, Ω
κ_i	wave number in coordinate direction x_i
κ	von Kármán constant
λ_i	wave length in coordinate direction x_i
μ	dynamic viscosity of the fluid, i.e. $\mu = \rho_f \nu$
ν	kinematic viscosity of the fluid
ω	frequency of time signal
ω_c	frequency used to define convection velocity of a given wave number (cf. equation 3.18)
ω_p	angular velocity of particle with respect to its centre of mass
Ω	control volume bounded by Γ
Ω_p	angular position of a particle with respect to Cartesian coordinate axes
ϕ	scalar variable, e.g. a component of the velocity vector
ϕ_s	solid volume fraction
ϕ_s^a	near-bed reference concentration (cf. equation 5.1)
ϕ_s^g	global solid volume fraction, i.e. $N_p^s \pi D^3/(6L_xL_yL_z)$
ψ	scalar variable, e.g. a component of force or torque on a particle
ρ_f	fluid density
ρ_p	particle density
τ_D	stress on particle surface contributing to drag
τ_L	stress on particle surface contributing to lift

τ_T	stress on particle surface contributing to spanwise torque
τ_{tot}	total mean shear stress in open and closed channel flow
τ_w	mean wall shear stress
$\boldsymbol{\tau}$	shear stress tensor, $\boldsymbol{\tau} = \nu\rho_f(\partial_j u_i + \partial_i u_j)$
τ_ℓ	integral time scale (cf. equation 3.15)
τ_m	micro time scale (cf. equation 3.16)
τ	time lag
τ_Δ	difference of time-lags related to maximum to minimum in cross-correlation (cf. table 3.11)
θ	azimuthal angle in wall-parallel plane (cf. equation 3.9)
θ_s	Shields parameter (cf. equation 1.1)
ξ	spatial shift in streamwise direction
ζ	spatial shift in spanwise direction

Roman symbols

A	constant in logarithmic law of the wall (cf. equation 3.1)
A_R	reference area, here defined as $A_R = L_x L_z / N_p$
A_{sph}	surface area of a sphere $A_{sph} = \pi D^2$
Ar	Archimedes number
B	constant in law of the wall utilising equivalent sand grain roughness k_s (cf. equation 3.2)
c_L	lift coefficient
c_F	coefficient related to definition of Shields number based on erosion due to translational motion (cf. §4.3)
c_T	coefficient related to definition of Shields number based on erosion due to rotation (cf. §4.3)
C	constant in law of the wall utilising equivalent sand grain roughness k_s (cf. equation 3.2)
C_{Fi}	force coefficient in x_i direction, i.e. $\langle F_i \rangle / F_R$
C_{Ti}	torque coefficient in x_i direction, i.e. $\langle T_i \rangle / T_R$
ds	area of a surface element
D	diameter of a sphere

D_r	diameter of rods
$\mathbf{e}_i$	unit vector in coordinate direction x_i
$\mathbf{f}$	volume force
F_R	reference force, here defined as $F_R = \rho_f u_\tau^2 A_R$ (cf. §3.3)
F_t	Force on a particle with angle α to the horizontal axis (cf. equation 4.1)
$\mathbf{F}$	force on a particle, $\mathbf{F} = (F_x, F_y, F_z)$
$\mathbf{F}^C$	force on a particle due to a contact model
$\mathbf{F}^\Gamma$	force on a particle due to surface stresses
$\mathbf{F}^\Omega$	force on a particle due to volume forces
$\mathscr{F}_i$	force on a square surface element in a smooth wall, $\mathscr{F}_i = (\mathscr{F}_x, \mathscr{F}_y, \mathscr{F}_z)$
g	gravitational constant in wall-normal direction
h	effective open channel height, i.e. $h = H - y_0$
H	height of computational domain
I_{bin}	intervals considered to compute Eulerian particle statistics
I_p	moment of inertia of a particle
I_t	time interval defined by the sequence a quantity ϕ exceeds a threshold ϕ_{thres} (cf. equation 4.3)
$\mathscr{I}$	integral of a quantity ϕ over the interval I_t defined by ϕ_{thres}
k_s	equivalent sand grain roughness
$k_{s\infty}$	equivalent sand grain roughness in fully rough regime
K_ϕ	kurtosis of a quantity ϕ
Kn	Knudsen number
L_x	length of domain in streamwise direction
L_y	length of domain in wall-normal direction
L_z	length of domain in spanwise direction
$\mathscr{L}$	characteristic length scale
m_p	mass of particle
n_s	number density of Eulerian particle statistics (cf. equation D.4)
$\mathbf{n}$	unit normal vector

N_{bin}	number of bins under consideration
N_{cond}	number of samples that meet averaging condition under consideration
N_I	number of intervals under consideration
N_t	number of time steps under consideration
N_p^s	number of mobile particles
N_p	number of particles in fixed particle layer
p	periodically varying part of total pressure
p_l	linearly varying part of total pressure
p^{tot}	pressure, i.e. $p^{tot} = p_l + p$
r_R	reference distance, here defined as $r_R = y_0 - D/2$
$\mathbf{r}_c$	position vector from contact points to particle centre
$\mathbf{r}_p$	position vector from particle centre to an element of the particles surface Γ
$\mathbf{r}_s$	position vector with respect to the centre of a square surface element
R	correlation function
Re	Reynolds number (cf. equation 2.4)
Re_b	bulk Reynolds number, here defined as $\text{Re}_b = U_{bH}H/\nu$
Re_τ	friction Reynolds number, here defined as $\text{Re}_\tau = u_\tau h/\nu$
s	side length of square surface element
S_ϕ	skewness of a quantity ϕ
$\mathscr{S}_\phi$	cumulative function of the stress contribution to mean value of drag ($\phi = \tau_D$), lift ($\phi = \tau_L$) and spanwise torque ($\phi = \tau_T$) on a particle
t	time
T_E	torque on a particle with respect to axis of downstream support (cf. equation 4.2)
T_R	reference torque, here defined as $T_R = F_R r_R$ (cf. §3.4)
$\mathbf{T}$	torque on a particle, $\mathbf{T} = (T_x, T_y, T_z)$
$\mathscr{T}$	period of time signal
$\mathscr{T}_y$	torque on a square surface element in a smooth wall
u	velocity component in streamwise direction
u_c	convection velocity of a quantity ϕ for a given wave number

u_τ	friction velocity
$\mathbf{u}$	velocity vector, $\mathbf{u} = (u, v, w)$
$\mathbf{u}_p$	velocity of particle with respect to its centre of mass
U_{bh}	bulk velocity based on h, i.e. $U_{bh} = 1/h \int_{y_0}^{H} \langle u \rangle \, \mathrm{d}y$
U_{bH}	bulk velocity based on the domain height, i.e. $U_{bH} = 1/H \int_0^H \langle u \rangle \, \mathrm{d}\, y$
U_c	convection velocity of a quantity ϕ
U_s	nominal settling velocity of a particle
$\mathscr{U}$	characteristic velocity scale
v	velocity component in wall-normal direction
V_p	volume of particle, i.e. $V_p = 1/6\pi D^3$ in case of a sphere
V_{fb}	total volume of fluid in a periodic box around a particle
w	velocity component in spanwise direction
x	Cartesian coordinate in streamwise direction
$\mathbf{x}$	position vector, i.e. in Cartesian coordinates $\mathbf{x} = (x, y, z)$
$\mathbf{x}_p$	position of the particle with respect to its centre of mass, i.e. $\mathbf{x}_p = (x_p, y_p, z_p)$
$\mathbf{x}_s$	position of centre of a square smooth wall surface element, i.e. $\mathbf{x}_s = (x_s, y_s, z_s)$
y	Cartesian coordinate in wall-normal direction
y_0	position of virtual wall
y_{bin}	centre of bins in wall-normal direction
y_{bin}^b	boundaries of bins in wall-normal direction
z	Cartesian coordinate in spanwise direction

Abbreviations

CPU	central processing unit
DNS	direct numerical simulation
LES	large eddy simulation
RANS	Reynolds averaged Navier–Stokes equation

Chapter 1

Introduction

1.1 Motivation

Turbulent open channel flow, sediment erosion and sediment transport is important to fluvial engineering applications. For example, sediment erosion around piers can lead to the collapse of bridges. Also, the capacity of water reservoirs and water power dams can be considerably reduced by sanding up of the basin upstream of hydraulic structures. Altering the sediment flux in rivers by weirs can lead to a deepening of the river bed downstream of weirs with severe consequences for the ground water level and the ecosystem in the vicinity of the river. Apart from these hydraulic applications, sediment transport is also of relevance to other fields of research. Geologists study the details of sediment transport and sediment erosion to understand the processes that shape landscapes and lead to specific types of rock formations. For some engineering applications, the transport of solid material by a carrier fluid can be highly beneficial. Details of sediment transport in pipe flow is of interest to these applications as well as to applications in the oil industry. There, the transport of solids can lead to blockage of the production pipes and reduce the productivity of wells.

Despite the importance of sediment erosion and sediment transport, the fundamental aspects involved are far from being completely understood. At the core of the problem is the complex interaction between a turbulent flow field with solid particles. The turbulent flow induces a hydraulic force and a hydraulic torque on a particle which can lead to sediment erosion and define its motion when suspended in the flow. In reverse, particles alter the structure of the turbulent flow when suspended or by posing a rough wall boundary condition when resting on the bed. Both phases, the fluid as well as the solid phase are related to a high degree of complexity. Turbulent flow in an open channel, such as flow in a river, is statistically in-homogeneous in at least one spatial direction, and the Reynolds numbers of interest are typically large, which leads to a wide range of velocity, length and time scales. Similarly, the solid phase can be related to a range of sediment sizes and shapes and the distribution of the phase is in general not trivial. Under the effect of gravity, sediment in river flow tend to accumulate near the bed of the channel and form a water worked bed. Such a bed is not necessarily randomly structured and can be described only by statistical means. The complexity is further increased when under certain conditions, the hydrodynamic force and torque induced by the turbulent flow erodes the bed and entrains particles. The particles are then transported by the fluid until they come to rest once more. In case of sediment transport with a constant sediment flux, the cycle of erosion and deposition leads to a concentration profile that decreases with the distance from the bed.

To improve the understanding of the fundamental aspects involved in sediment erosion and sediment transport, it appears necessary to simplify the problem, while, at the same time, retaining some of the fundamental physics. This study approaches the topic by increasing the complexity of the problem in question in three steps. First, turbulent flow over a fixed bed of mono-sized spheres in a structured arrangement is investigated. Special emphasis is on the characterisation of hydrodynamic force and torque on particles as well as on the flow structures related to the force and torque. In the next step, the onset of sediment erosion is studied by simulations with fixed and mobile spheres. Focus is given on the flow structures that relate to sediment erosion and the characterisation of the erosion process. In the last step, the problem of sediment transport in an open channel with a constant sediment flux is considered. Here, turbulent flow that leads to erosion and sedimentation of particles is altered by the motion of the spheres. The differences to single-phase flow are studied in detail.

High-fidelity data with a high resolution in space and time of the flow configurations described above is generated by direct numerical simulation employing an immersed boundary method to resolve the particle–fluid interface. A brief literature review of the most relevant studies will be given in the following.

1.2 Literature review

1.2.1 Flow over rough walls

A large body of literature exists that deals with the characteristics of flow over rough surfaces. A reference for the earlier work on roughness is Schlichting (1965); a more recent review on the subject, including numerical studies is given by Jiménez (2004). Some consequences of roughness for high Reynolds number experiments have recently been reviewed by Marusic *et al.* (2010). Some of the key questions of interest in the research on rough wall turbulence are how roughness influences the turbulence structure, what are the consequences for scaling, and how can the effect on the fluid be estimated from the roughness geometry. Thus, the focus is almost exclusively on the effect of the rough wall on the fluid and the nature of the fluctuating force and torque acting on individual roughness elements is mostly not investigated.

Numerical studies of rough wall flows are very demanding in terms of computing time, much more so than comparable simulations of flow over smooth walls. Direct numerical simulation of flow over a wavy wall have been carried out by De Angelis *et al.* (1997), Cherukat *et al.* (1998) and more recent by Yang & Shen (2010). Direct numerical simulation of channel flow over a wall roughened by spanwise-oriented square bars have been carried out by Leonardi *et al.* (2003, 2007) and Ikeda & Durbin (2007). Turbulent flow over spanwise oriented square bars with heat transfer was investigated by Miyake *et al.* (2001) and Nagano *et al.* (2004). Orlandi & Leonardi (2008) have simulated plane channel flow including different layouts of wall-mounted cubes. Direct numerical simulations of channel flow with wall velocity disturbances (acting as artificial roughness) have been carried out by Orlandi *et al.* (2003) and Flores & Jiménez (2006). More in line with the present setup, Singh, Sandham & Williams (2007) have performed simulations of open channel flow over spheres in hexagonal arrangement, albeit at considerably coarser resolution than the one employed in the present study.

1.2.2 Hydrodynamic force on fixed particles

Several publications focus on the hydrodynamic force acting on spherical objects placed in a fluid flow. In the low Reynolds number range analytical solutions have been proposed for various flow configurations, e.g. the case of a particle in a linear shear flow (Saffman, 1965; Auton, 1987), in a non-uniform rotational flow (Auton, Hunt & Prud'homme, 1988), and of a particle in the vicinity of a smooth wall (Krishnan & Leighton, 1995). In order to gain information on the mechanism that leads to lift and drag on a particle in the range from small to moderate Reynolds numbers, similar flow configurations have also been explored by means of experimental studies (King & Leighton, 1997) and by means of direct numerical simulations (Kim *et al.*, 1993; Bagchi & Balachandar, 2002; Zeng *et al.*, 2009; Lee & Balachandar, 2010). In the high Reynolds number limit, numerous studies can be found that describe the flow around spheres in unbounded flow (see Yun, Kim & Choi, 2006, for an overview).

The studies mentioned above have focused on situations in which the flow field approaching the sphere is laminar in nature. It is well known, however, that turbulence can have a significant effect on the statistics of the force acting on a sphere. A review on the effect of turbulence on an isolated sphere can be found in Bagchi & Balachandar (2003). The authors studied the forces on an isolated sphere subject to free-stream isotropic turbulence for small and moderate Reynolds numbers by means of direct numerical simulation. They found that turbulence had only little effect on the mean drag and that the fluctuations of lift and drag scaled linearly with both the mean drag and the turbulence intensity.

In contrast, turbulence appears to have a significant effect in the case of a sphere positioned close to a wall, the mean lift being particularly affected (Willetts & Murray, 1981; Hall, 1988; Zeng *et al.*, 2008). The experimental evidence shows, that similar to the low Reynolds number regime, significant positive values for mean lift (directed away from the wall) are obtained for a sphere touching the wall plane (Willetts & Murray, 1981; Hall, 1988; Mollinger & Nieuwstadt, 1996; Muthanna *et al.*, 2005). When the sphere is not touching the wall, the picture is less clear and still a matter of discussion: both positive and negative values of the lift are reported. Willetts & Murray (1981) found changes in sign for the value of the mean lift when increasing the wall distance; Hall (1988) measured consistently positive values for various wall distances; Zeng *et al.* (2008) obtained negative values (directed towards the wall) in case the sphere is placed in the buffer layer. Zeng *et al.* (2008) note that the classical formulae based on unbounded shear flow fail to predict the results from their direct numerical simulation correctly, stating that further investigations are required to understand the discrepancy.

Hall (1988) showed that the effect of a nearby wall on the lift experienced by a spherical body differs significantly depending on the wall being rough or smooth. In particular, it was found that the lift significantly decreased when the sphere was positioned in between spanwise oriented, rod-shaped roughness elements. When the sphere was positioned on top of the array of wall-mounted rods, however, the measured lift was comparable to the corresponding smooth wall values.

The difficulties related to the direct measurement of particle forces as in the studies above have been discussed by Muthanna *et al.* (2005). Another approach was taken by Einstein & El-Samni (1949). They approximated the force exerted on hemispheres in an open channel flow by pressure measurements on top and near the bottom of the hemispheres. They reported positive lift on the hemispheres, and were among the first who stated the relevance of the forces on particles in a rough wall to the understanding of sediment erosion. More recent studies following this approach present approximations of lift and drag on cubes, spheres and naturally shaped stones by local pressure measurements (Hofland, Battjes & Booij, 2005; Hofland & Battjes, 2006; Detert, Weitbrecht & Jirka, 2010*b*). These studies have focused on the higher Reynolds number regime with particle Reynolds numbers of the order of thousands.

1.2.3 Onset of sediment erosion

The start of modern research on the onset of sediment erosion is commonly linked to the doctoral thesis of Shields (1936). He proposed to define the onset of sediment erosion by the parameter that now bears his name. The Shields parameter is defined as

$$\theta_s = \frac{\tau_w}{(\rho_p - \rho_f)\, g D}, \tag{1.1}$$

where ρ_p is the density of the sediment particles, ρ_f is the density of the fluid, g the value of the gravitational acceleration and τ_w the wall shear-stress which needs to be further defined in the context of a geometrically rough wall. Shields suggested that the value of θ_s related to the onset of erosion, i.e. θ_{thres}, is a function of the particle Reynolds number, Du_τ/ν, where $u_\tau = \sqrt{\tau_w/\rho_f}$ is the friction velocity, D is the particle diameter and ν the kinematic viscosity. The doctoral thesis remained the only contribution of Shields to the subject, albeit with a strong impact on the hydraulic community, which lead to that his work and also his life have been a matter of various discussions (see Kennedy, 1995, and Buffington, 1999, for an overview).

Several studies related to sediment erosion follow the approach of Shields. A review of the literature, that can also serve as an introductory text, can be found in manuals and textbooks on the subject, e.g. Vanoni (1975), pp. 91; Yalin (1977), §4; van Rijn (1993), §4; García (2008), pp. 44. A compilation and discussion of experiments in the style of Shields is given by Buffington & Montgomery (1997). Some recent studies focused on the onset of sediment erosion in low Reynolds numbers (e.g. Charru *et al.*, 2007; Ouriemi *et al.*, 2007; Lobkovsky *et al.*, 2008; Peysson *et al.*, 2009). It should be noted, that a good approximation to the data of Shields can be derived from assuming a relation between mean drag on a exposed particle and the law of the wall (Ikeda, 1982; García, 2008).

Despite the wide spread used of Shield's approach to define the onset of sediment erosion, the scatter of θ_{thres} obtained in experiments remains unsatisfactorily large. This might be related to challenges inherent to the problem of defining the onset of erosion (see for example Naden, 1987 for a discussion). As has been argued in the beginning, the geometrical properties of the bed is in general not trivial due to the wide range of particle size, shape, densities and the complexity related to particle arrangements. Being additionally linked to a highly turbulent flow field the characteristics of force and torque on particles are even more complex and might be considered as random variables. Some authors argue that when described as a function of a random variable the onset of sediment erosion has a certain probability which can never be precisely zero (García, 2008, p. 46, Papanicolaou *et al.*, 2002). In addition to this probabilistic aspect, the definition of what is considered as erosion of a particle cannot be made unambiguously. For example, the motion of a particle across a bed is often classified into saltation and rolling, where the latter might not be considered as erosion. Also, Charru *et al.* (2007) found that the mean distance of a particle moving over an array of spheres from one resting position to the next is related to the Shields number. Thus, the definition of what is considered as erosion of a particle might influence the value of the critical Shields number.

Realising the limitations of defining the onset of erosion by a single global parameter, some studies concentrate on the flow structures that relate to high force events on fixed particles or the flow structures that lead to the onset of sediment erosion. Sutherland (1967) was among the first to investigate flow structures related to the onset of sediment erosion. More recent studies employing truly mobile particles are those by Niño & García (1996), Niño *et al.* (2003), Hofland (2005), Cameron (2006), Dwivedi (2010) and Dwivedi *et al.* (2010). Experiments on flow structures related to possible sediment erosion predicted based on fixed particle data have been carried out by Detert (2008), Detert *et al.* (2010*a*), Dwivedi (2010) and Dwivedi *et al.* (2010). The results of the recent experiments based on fixed and mobile particles indicate that sediment erosion is often related to regions of positive streamwise velocity fluctuations in the vicinity of the eroding particle.

1.2.4 Sediment transport

Performing experiments in multi-phase flow is difficult as the solid phase can interfere with the measurement device. In spite of these difficulties, several configurations related to particle laden open channel flow were investigated experimentally over the last two decades (Kaftori *et al.*, 1995; Taniere *et al.*, 1997; Kiger & Pan, 2002; Righetti & Romano, 2004*b*; Muste *et al.*, 2009). Most of these experiments were performed using dilute suspensions, with a volumetric concentration of order $\mathcal{O}(10^{-3})$, and various kinds of sediments. For example, in the study of Taniere *et al.* (1997) the density ratio between the solid and fluid phase was very high since they employed glass and polymer particles in air. The density ratio was lower in the rest of the studies, ranging from values as low as 1.05 in the experiments of Kaftori *et al.* (1995) (polystyrene particles in water) to 2.6 in Kiger & Pan (2002) and Righetti & Romano (2004*a*) (both used glass particles in water), while Muste *et al.* (2009) covered the range 1.03 to 2.65 using natural sand and neutrally-buoyant sand (crushed nylon) in water. These experiments have shown the existence of a velocity lag (in the mean) between particle and fluid velocities, and modifications to the turbulence characteristics of the flow due to the presence of the suspended particles.

Computational studies are only now beginning to appear. An exception is the pioneering work of Pan & Banerjee (1997), who conducted resolved direct numerical simulations of turbulent particulate flow in a horizontal channel using 160 stationary and mobile particles.

1.3 Objectives and structure of this thesis

The objective of this thesis is to first provide and then analyse high-fidelity data of (i) turbulent open channel flow over an array of fixed spheres, of (ii) the erosion process of mobile spheres in this arrangement and of (iii) turbulent open channel flow with transport of many moving spheres in a cycle of re-suspension and sedimentation. The data analysis aims to close knowledge gabs related to force and torque on fixed particles, answer open questions with respect to the onset of sediment erosion and contribute to the understanding of sediment transport. The data is obtained with a high resolution in space and time by employing direct numerical simulations (DNS) with an immersed boundary method to fully resolve the fluid-particle interface. To ascertain confidence in the simulations, emphasis is given on comparing the results of the simulations to results in the literature when possible.

The thesis is structured as follows. Chapter 2 provides an overview on the governing equations and fundamental concepts. Chapter 3 presents the results of two DNSs of turbulent open channel flow over fixed mono-sized spherical particles in a structured arrangement. Characteristics of the flow field, the statistics of force and torque on a particle as well as characteristics of the flow structures related to force and torque are discussed in detail. In chapter 4 the implications of the fixed sphere results for the onset of sediment erosion is studied. Furthermore, simulations of the onset of erosion with mobile particles are analysed. In chapter 5 the results of four DNSs of horizontal, open channel flow with sediment transport of many mobile particles are reported. Focus is given on the statistics of particles and flow field. Each chapter begins with a short introduction and ends with a summary, conclusions and recommendation for future work. A summary, conclusions and recommendation for future work of the thesis is given in chapter 6. Some additional material and extended discussion is provided in the appendices.

Chapter 2

Fundamentals

Grau, teurer Freund, ist alle Theorie,
und grün des Lebens goldner Baum.

Johann Wolfgang Goethe

This chapter provides a brief summary of some fundamental aspects related to this thesis. On one hand, it serves as a reference for the following chapters, on the other hand it provides references for further reading. Section 2.1 provides a review on the governing equations of fluid flow (§2.1.1), solid body motion (§2.1.2) and the coupled fluid-solid system (§2.1.3). An overview on the phenomenon of turbulence is given (§2.2.1), followed by an introduction to numerically studying turbulent flow (§2.2.2). A description of the numerical method used in this thesis and some validation aspects is given in §2.2.3.

2.1 Governing equations

2.1.1 Fluid phase

2.1.1.1 Continuum hypothesis and Navier–Stokes equation

In the process of deriving a mathematical description of fluid flow several assumptions are made. One of the most fundamental assumptions is the treatment of fluids as a continuum, i.e. that its macroscopic behaviour can be described by being continuous in space. Thus the molecular structure of the fluid is neglected and the effect of it is taken into account only in an average sense. A measure for when the continuum hypothesis can be considered appropriate is the Knudsen[1] number, Kn, which is defined as the ratio of a characteristic molecular length scale, e.g. the mean free path of molecules, and a characteristic length scale of macroscopic flow properties, e.g. a length scale of the smallest scales of fluid motion. In most engineering applications is it found that $\mathrm{Kn} \ll 1$ by several orders of magnitudes, which indicates that the continuum hypothesis holds (cf. Batchelor, 1967, §1.2; Pope, 2000, §2.3).

Other assumptions concern the variations of the fluids density, ρ_f, and the relation of shear stress to rate-of-strain. In the present study only incompressible fluids are considered, that is the density of the fluid is independent of space and time, $\rho_f=$ const. Furthermore, the fluid is considered to be

The biographical information on people in honour of whom numbers, formulae or hypothesis are named is based on the information provided in the internet encyclopedia Wikipedia in fall 2011.

[1] Martin Hans Christian Knudsen, Danish physicist, ⋆ 15 February 1871 † 27 May 1949

Newtonian[1], that is the relation between the non-isotropic part of the shear stress and the rate-of-strain tensor is linear (cf. Batchelor, 1967, §3.3, p. 147)

$$\boldsymbol{\tau} = \mu \left(\frac{\partial u_i}{\partial x_j} + \frac{\partial u_j}{\partial x_i} \right), \qquad \text{for } i,j = 1,2,3, \tag{2.1}$$

where $\boldsymbol{\tau}$ is the viscous stress tensor, μ is the dynamic viscosity of the fluid defined as the product of the density of the fluid, ρ_f, with the kinematic viscosity of the fluid ν, $\mu = \rho_f \nu$, $\mathbf{u}$ is the velocity vector $\mathbf{u} = (u, v, w)$ in Cartesian coordinates with axes x, y and z in streamwise, vertical and spanwise direction respectively. Note, that in the following bold symbols are used to denote tensor or vector quantities and indices 1, 2, 3 or x, y, z to denote vector components in the Cartesian reference frame, e.g. f_x or f_1 denote the streamwise component of a vector $\mathbf{f}$.

Based on the principle of conservation of mass, the continuity equation for an incompressible fluid reads

$$\nabla \cdot \mathbf{u} = 0, \tag{2.2}$$

where ∇ is the Nabla operator, i.e. in Cartesian coordinates $\nabla = (\partial/\partial x, \partial/\partial y, \partial/\partial z)$, and $\cdot$ denotes the dot product between two vectors.

Based on the principle of conservation of momentum, the momentum equation for an incompressible Newtonian fluid can be written as

$$\frac{\partial \mathbf{u}}{\partial t} + (\mathbf{u} \cdot \nabla)\mathbf{u} = -\frac{1}{\rho_f}\nabla p + \nu \nabla^2 \mathbf{u} + \mathbf{f}, \tag{2.3}$$

with time t, pressure p, Laplace[2] operator, $\nabla^2 = \nabla \cdot \nabla$, and a volume force, $\mathbf{f}$, on the fluid.

The set of equations (2.2) and (2.3) are called Navier[3]–Stokes[4] equations in the following. A detailed derivation of the Navier–Stokes equations can be found for example in Schlichting (1965), Batchelor (1967) or Pope (2000)

With reference scales for length, time, velocity and pressure the Navier–Stokes equation (2.2) and (2.3) can be written in dimensionless form. This shows that only a single non-dimensionless parameter governs the characteristics of the Navier–Stokes equations, namely the Reynolds[5] number, defined as

$$\mathrm{Re} = \mathscr{U}\mathscr{L}/\nu, \tag{2.4}$$

where $\mathscr{U}$ and $\mathscr{L}$ are a characteristic velocity and length scale of the flow, respectively. The Reynolds number represents the ratio of convective terms to viscous terms, i.e. for small values of Re the viscous terms dominate the fluid motion and the flow is commonly found to be laminar. For large values of Re the convective terms dominate the fluid motion and the flow is commonly found to be turbulent.

2.1.1.2 Boundary conditions and initial conditions

The Navier–Stokes equations (cf. equations 2.2 and 2.3) are a set of four partial differential equations – one continuity equation and one momentum equation in each coordinate direction – for four unknown variables, i.e. u, v, w and p. For a given set of appropriate boundary and initial conditions the evolution

[1] Sir Isaac Newton, English polymath, ⋆ 4 January 1643 † 31 March 1727

[2] Pierre-Simon, marquis de Laplace, French mathematician and astronomer, ⋆ 23 March 1749 † 5 March 1827

[3] Claude-Louis Navier, French engineer and physicist, ⋆ 10 February 1785 † 21 August 1836

[4] Sir George Gabriel Stokes, Irish mathematician and physicist, ⋆ 13 August 1819 † 1 February 1903

[5] Osborne Reynolds, British scientist and engineer, ⋆ 23 August 1842 † 21 February 1912)

of a flow field in time can be obtained by integration; initial conditions for the flow variables need to be specified in the entire domain Ω at the initial time, boundary conditions need to be specified at all times along the boundary Γ of the domain Ω under consideration. Boundary conditions of relevance to this thesis are reviewed below, followed by a discussion on some aspects of the initial condition.

Dirichlet and Neumann boundary conditions A boundary condition for a variable, ϕ, can be to specify its value for a given position $\mathbf{x}$ in time, t, i.e. $\phi(\mathbf{x},t) = c$, where $\mathbf{x}$ is located on the domain boundary Γ and c is a certain value. Such a boundary condition is called Dirichlet[1] boundary condition. An example of a Dirichlet boundary condition is a stationary plane wall at $y = 0$. In case of a no-slip condition for the fluid at the wall, i.e. at the boundary the velocity of the fluid on the boundary is equal to the velocity of the boundary at each point, the velocity boundary condition at all times can be written as

$$\mathbf{u}(x,0,z,t) = (0,0,0)\,. \tag{2.5}$$

Another possible boundary condition is to specify the gradient of a variable, e.g. $\partial\phi/\partial\mathbf{n} = c$, where $\mathbf{n}$ is the normal vector of Γ. Such a boundary condition is called Neumann[2] boundary condition. An example for a Neumann boundary condition is the boundary condition for pressure at domain boundaries of a computational domain which is often defined as zero

$$\frac{\partial p}{\partial \mathbf{n}} = 0\,. \tag{2.6}$$

Periodic boundary conditions In turbulence research the domain is often considered to be unbounded in one or more directions. For example, in the case of open or closed channel flow the domain is considered infinite in streamwise and spanwise direction. However, infinite domain dimensions can lead to mathematical difficulties (Frisch, 1995, §2.1, p. 14) and, above all, to numerical difficulties. An approach to simplify matters is to consider the flow field to be periodic in the respective directions. In the example of channel flow with a wall-normal (y) extension L_y and periodic boundary conditions in streamwise (x) and spanwise (z) direction with periodicities L_x and L_z this can be written as

$$\begin{aligned} \mathbf{u}(x,y,z,t) &= \mathbf{u}(x+nL_x,y,z+mL_z,t)\,, \\ p(x,y,z,t) &= p(x+nL_x,y,z+mL_z,t)\,, \end{aligned} \tag{2.7}$$

for any integer n, m and $0 \le x < L_x$, $0 \le y < L_y$, $0 \le z < L_z$. The case of a non-periodic infinite flow field can be recovered by letting the periodicity go to infinity, i.e. $L_x \to \infty$ and $L_z \to \infty$. Thus, for large ratios L_x/L_y and L_z/L_y, the effect of the periodic boundary condition with respect to the infinite solution to the problem might be considered small. However, for small ratio L_x/L_y and L_z/L_y, the effect can be large and influence the results considerably. This is because the largest scales of turbulent motion in the periodic direction are limited to the length of periodicity. Some studies make use of this aspect, to isolate certain scales of turbulent motion (Jiménez & Moin, 1991, Flores & Jiménez, 2010).

Free-slip boundary condition and rigid-lid assumption A boundary condition of relevance to the present work is the so called free-slip boundary condition. This boundary condition consists of a

[1]Johann Peter Gustav Lejeune Dirichlet, German mathematician, ⋆ 13 February 1805 † 5 May 1859

[2]Carl Gottfried Neumann, German mathematician, ⋆ 7 May 1832 † 27 March 1925

zero-velocity condition for the velocity component normal to the boundary, and a condition of zero-gradient for the other velocity components. For example, a free-slip condition at location $y = H$ can be specified via

$$\frac{\partial u(x,H,z,t)}{\partial y} = 0, \quad v(x,H,z,t) = 0, \quad \frac{\partial w(x,H,z,t)}{\partial y} = 0. \tag{2.8}$$

Although such a boundary condition is difficult to realize experimentally, it is of relevance to the simulation of open channel flow. In an open channel flow an interface exists, that separates lighter fluid, e.g. air, from a heavier fluid, e.g. water. The interface varies in height with time and thus is not fixed at a certain position. Assuming that the surface deformation is small, the variation of the surface might be neglected. This is called rigid-lid assumption and is equivalent to impose a free-slip condition for the fluid velocity as described above. A more detailed discussion of the rigid-lid assumption can be found in Komori *et al.* (1993).

Initial conditions For steady boundary conditions, that is the boundary conditions are constant in time, the flow can reach a state in which its statistics become independent of time, although, the flow field itself might be highly time dependent. The flow is then called to be in a statistically stationary state. It is generally assumed that for a given flow configuration the same statistically stationary state is reached – independently from the initial conditions. Thus the specific definition of initial conditions is often of minor interest when the characteristics of the statistically stationary state are considered.

In numerical simulations the integration of the flow field in time until a statistically stationary state is reached can be very costly. The time needed is influenced strongly by the flow configuration under consideration and the choice of the initial conditions. It can be beneficial to specify the initial condition by interpolation of a flow field from a coarser simulation that reached a statistically stationary state. In a coarser simulation, the computational costs to overcome initial effects can be considerably smaller.

2.1.2 Solid phase

2.1.2.1 Equations of rigid-body motion

Similarly to fluids, solids can be described as continua. This is of particular interest when the deformation of solids or the stress distribution within solid are important. Here, the deformation of solid particles due to hydrodynamic stress is much smaller than the particle dimension and is therefore neglected. Instead of a continuum approach, the particles are described by the mechanics of a rigid-body. A rigid-body can be defined as a solid for which the distance between two points of the solid does not change in time. This corresponds to assuming that the modulus of elasticity approaches infinity.

The equations of rigid-body motion can be derived from Newton's second law for linear and angular momentum (cf. Goldstein *et al.*, 2002). For spherical particles these read

$$\frac{\mathrm{d}\mathbf{x}_p}{\mathrm{d}t} = \mathbf{u}_p, \qquad m_p \frac{\mathrm{d}\mathbf{u}_p}{\mathrm{d}t} = \mathbf{F}, \tag{2.9}$$

$$\frac{\mathrm{d}\boldsymbol{\Omega}_p}{\mathrm{d}t} = \boldsymbol{\omega}_p, \qquad I_p \frac{\mathrm{d}\boldsymbol{\omega}_p}{\mathrm{d}t} = \mathbf{T}, \tag{2.10}$$

where the position of a particle is $\mathbf{x}_p$, the velocity of a particle is $\mathbf{u}_p$, the force on the particle is $\mathbf{F}$, the mass of the particle is m_p, the angular position of a particle with respect to the Coordinates axes is

$\boldsymbol{\Omega}_p$, the angular velocity of the particle is ω_p, the torque on a particle is $\mathbf{T}$ and the moment of inertia is I_p. The quantities above are defined with respect to the centre of mass of the spherical particle. Note, that for arbitrarily shaped rigid bodies the moment of inertia is a tensor quantity not a scalar as denoted above for the case of spherical particles. For arbitrarily shaped bodies the time derivation in (2.10) would need to consider the product of moment of inertia tensor and angular velocity. The equations (2.9) and (2.10) can be solved for the four quantities $\mathbf{x}_p$, $\mathbf{u}_p$, $\boldsymbol{\Omega}_p$ and ω_p when appropriate initial and boundary conditions are specified, i.e. the initial conditions for $\mathbf{x}_p$, $\mathbf{u}_p$, $\boldsymbol{\Omega}_p$ and ω_p need to be specified as well as the force and torque on the particle at each time as boundary conditions.

2.1.2.2 Force on a particle

In this thesis, the force on a particle, $\mathbf{F}$, is considered as the sum of (i) the hydrodynamic force, $\mathbf{F}^\Gamma$, defined as the integral of hydrodynamic stress on the particle surface, (ii) the body force, $\mathbf{F}^\Omega$, due to the acceleration of the particle by a volume force such as gravity and (iii) the model force, $\mathbf{F}^C$, to account for particle–particle or particle–wall contact. Each of these forces are described in more detail below.

Hydrodynamic force Hydrodynamic stresses on the surface of an impermeable particle in a fluid arise as a result of the no-slip condition between the fluid velocity and the velocity at the particle surface, i.e. the velocity of the fluid and particle surface are identical at every point of the particles surface. The hydrodynamic force can be defined as

$$\mathbf{F}^\Gamma = \int_\Gamma \boldsymbol{\tau} \cdot \mathbf{n} \, \mathrm{d}\Gamma - \int_\Gamma p^{tot} \mathbf{n} \, \mathrm{d}\Gamma, \tag{2.11}$$

where Γ denotes the particle-fluid interface and p^{tot} denotes the hydrodynamic pressure.

Body force The resulting force on a particle in a constant density fluid as a result of a volume force, $\mathbf{f}$, acting on the fluid as well as on the particle, can be expressed as

$$\mathbf{F}^\Omega = \int_\Omega (\rho_p - \rho_f) \mathbf{f} \, \mathrm{d}\Omega, \tag{2.12}$$

where ρ_p is the density of the solid particle and Ω denotes the volume of the particle. In case $\mathbf{f}, \rho_p$ and ρ_f are constant (2.12) simplifies to

$$\mathbf{F}^\Omega = (\rho_p - \rho_f) V_p \mathbf{f}, \tag{2.13}$$

where V_p is the volume of the particle, i.e. $V_p = \int_\Omega \mathrm{d}\Omega = 1/6\pi D^3$ in case of a spherical particle with diameter D.

Contact model force To account for the particle–particle and particle–wall contact a model is employed in the present study. The contact model is based on the artificial repulsion potential of Glowinski *et al.* (1999), relying upon a short-range repulsion force (here with a range of twice the grid width, i.e. $2\Delta_x$). Its primary use is to prevent particles of intersecting each other. The contact model has been applied in the related simulations of the fluidisation of spheres (Pan *et al.*, 2002) for which very good agreement was found to experiments, even in a rather dense configuration. The model was also successfully applied in simulations of particles transported in a vertical channel (Uhlmann, 2008). It should be noted that this model is rather crude and a more elaborated model might be favourable for some configurations. A more detailed discussion on the aspect is provided in the respective sections of this thesis. More details on contact mechanics can be found in textbooks such as Johnson (2003) or in the work of Clift *et al.* (1978); Crowe *et al.* (1998); Jackson (2000) and Prosperetti & Tryggvason (2007).

2.1.2.3 Torque on a particle

Similar to the forces discussed above, torque on a particle can be due to hydrodynamic stresses, volume forces and a model torque.

The torque on a spherical particle as a result of hydrodynamic stresses on its surface is defined as

$$\mathbf{T} = \int_{\Gamma} \mathbf{r}_p \times (\boldsymbol{\tau} \cdot \mathbf{n}) \; \mathrm{d}\Gamma, \tag{2.14}$$

where $\mathbf{r}_p$ is the distance vector from the particle centre to an element of the particles surface Γ, and $\times$ denotes the cross product of two vectors. Contrary to the definition of the hydrodynamical force on a particle (2.11), the pressure does not enter the integral (2.14), since in the case of spherical particles the differential pressure force $-p^{tot}\mathbf{n}\mathrm{d}s$ on a surface element $\mathrm{d}s$, is always directed towards the centre of the sphere. This thesis is limited to consider force and torque on spherical particles with constant densities under the influence of constant volume forces. Hence, the volume force does not contribute to torque on the particles. Also, the contact model does not consider tangential contact forces, and thus does not contribute to torque on a particle. In conclusion, the torque on a particle in this thesis is only due to relation (2.14).

2.1.3 Fluid – solid systems

Fluid flow with submerged mobile rigid bodies can be described by the two sets of equations discussed in sections 2.1.1 and 2.1.2. The coupling of the equations is established over the boundary conditions. A no-slip condition at the fluid–solid interface requires the velocity of the two phases to be equal and thus poses a boundary condition for the fluid. Similarly hydrodynamic force and torque on the particle surface jointly with body force and model force pose boundary conditions for the rigid particle motion. With the requirement, that all conditions need to be satisfied at each instant in time, the coupling between the two sets is established and the problem is well defined.

2.2 Numerical simulation of turbulent flow

In the previous section, the governing equations of fluid flow have been introduced. The present section gives a summary of the phenomenon of turbulence that can emerge from these equations, of the available tools to numerically study turbulent flow of a single-phase fluid as well as a brief description of the numerical method employed in this thesis to simulate fluid flow, as well as fluid flow with mobile rigid bodies.

2.2.1 Turbulence in fluids

Turbulence is a complex phenomenon that has been studied intensively for several decades in various fields of research, including mathematics, physics, engineering and biology. Often the questions related to turbulence differ from field to field, which might be one of the reasons, why a common definition of turbulence does not exist.

Turbulence in fluids is often introduced as a state of the flow, being either laminar or turbulent. The laminar state is characterised by a smooth variation of the flow variables and well defined flow characteristics. Following the ideas of Tennekes & Lumley (1972), the state of turbulent flow can be characterised by a large range of scales of irregular, diffusive, dissipative and highly three-dimensional motions. Furthermore the state of flow is related to large Reynolds numbers. Despite this somewhat loose definition, it appears to be sufficient to describe the turbulent flow. Examples to observe turbulent flow in every day life are cumulus clouds, smoke rising from a chimney, the mixing of milk in the morning tea or flames of a bonfire.

It is not hard to believe that the beauty of turbulent flow has fascinated mankind from the beginning. Indeed, some authors see the importance of turbulence to the hydraulical structures in the ancient civilisation as an indication that the responsible engineers must have had some understanding of turbulence and its effects (e.g. Rouse & Ince, 1957; Jiménez, 2000). One of the oldest documents on the study of turbulent flow dates back to the work of Leonardi Da Vinci[1] (e.g. book cover Tennekes & Lumley, 1972; Frisch, 1995, p. 112 and p. 183). In spite of these references, the advent of modern turbulence research is commonly related to the mid 19th century, when detailed measurements of the pressure drops in turbulent pipe flow were carried out by Hagen (1854) and Darcy (1857). Two decades later, Boussinesq (1877) suggested a statistical treatment of turbulent flows and introduced the concept of an enhanced eddy viscosity. A detailed study on the transition from the laminar to the turbulent flow regime, was carried out by Reynolds (1883) who proposed the criterion that now bears his name. Later, Reynolds (1895) introduced the concept of flow decomposition. With the beginning of the 20th century it became more and more clear how a phenomenon such as turbulence can emerge from a set of deterministic equations and it is now commonly accepted, that the Navier–Stokes equations are appropriate to describe turbulent flow. A major step in understanding the nature of turbulence was made by the work of A. N. Kolmogorov[2] (e.g. Kolmogorov, 1941). In his work Kolmogorov advanced and quantified the energy cascade concept described by Richardson (1922). According to the concept, turbulent energy is injected in the large scales and transferred to smaller scales by an inviscid process. The Reynolds number of the smallest scales is low, such that viscosity becomes important and dissipates all energy in the small scales. Thus, this model provides an explanation of the way the rate of dissipation might be specified by the inviscid processes related to the large scales and not by the viscosity itself. A review of the work of Kolmogorov can be found in the book by Frisch (1995).

The above introduction on the phenomenon of turbulence is far from being complete. A more detailed introduction can be found in review articles such as the one by Jiménez (2000) or more detailed in form of textbooks on the subject (for example in Tennekes & Lumley, 1972, Pope, 2000 or Davidson, 2007).

2.2.2 Direct numerical simulation of turbulent flow

Many textbooks classify the numerical approaches to study turbulent flow in direct numerical simulation (DNS), large eddy simulation (LES) and Reynolds averaged Navier–Stokes equations (RANS). These definitions are made according to the governing equation in each approach which are the Navier–Stokes equations, the filtered Navier–Stokes equations or the time-averaged Navier–Stokes equations, respectively. An overview on the three approaches can be found for example in the textbook by Pope (2000). The present section provides an overview on DNS, which is the method used to study the flow configurations in this thesis.

[1] Leonardo di ser Piero da Vinci, Italian polymath, ⋆ 15 April 1452 † 2 May 1519

[2] Andrey Nikolaevich Kolmogorov, Soviet Russian mathematician, ⋆ 25 April 1903 † 20 October 1987

It is now generally accepted that, although being deterministic and based on a continuum approach, the Navier–Stokes equations (cf. equations 2.2 and 2.3) are appropriate to describe turbulent flow. Therefore, the simplest numerical approach to study turbulent flow is to solve the Navier–Stokes equations without further modelling assumptions for the flow variables by integration in time. This requires, that all scales of the turbulent flow are resolved jointly with appropriate boundary and initial conditions. In particular, the largest scales need to be resolved by considering domains large enough to minimise boundary effects as well at a resolution fine enough to simulate the dissipation of energy at the smallest scales. It is important to note, that the scales of turbulence have to be resolved not only in space but also in time. That is, the time discretisation needs to be fine enough to resolve the smallest time scales in the flow while at the same time the flow needs to evolve long enough to assess the influence of the largest time scales.

The duration of a simulation depends on several aspects. When the statistically stationary state of a flow configuration is of interest, the flow field must evolve long enough from its initial state for the initials effects to be small (cf. section §2.1.1.2). Once the statistically stationary state of a flow configuration is reach, the flow field often needs to be advanced in time over a certain period to obtain converged statistics. The duration of the latter depends on the statistics of interest, on the flow configuration and also on the discretisation of the flow field. For example, the number of samples of one-dimensional flow statistics in open channel flow can be increased by additionally averaging over the homogeneous direction. Therefore, the needed observation time of the simulation might be small for large computational domains.

The requirements regarding the resolution in time and space and the duration to obtain converged statistics make DNSs computationally expensive. For example, the total number of grid points to meet the required spatial resolution in homogeneous isotropic turbulence is approximately proportional to the square of Reynolds number, Re, (i.e. more precisely $\mathrm{Re}^{9/4}$ Pope, 2000, pp. 347). The computational effort scales with Re^3 when the efforts related with the required duration to collect statistics are included. This shows, that the costs of DNS increases rapidly with the Reynolds number. The major limitation related to study turbulent flow phenomena by DNS comes from the amount of computational expenses one is willing or able to spend. Despite the immense costs, latest DNS for channel flow (Hoyas & Jiménez, 2006) or for boundary layer (Jiménez *et al.*, 2010; Schlatter & Örlü, 2010), reached Reynolds numbers that can be considered moderate. The total costs of each of those simulation is immense and of the order of several millions of CPU hours. Commonly, these simulations require a run-time of several months up to a year on high performance computers. The reward of these efforts is a wealth of data which is difficult, if not impossible, to obtain otherwise.

Since the papers of Kim *et al.* (1987) and Spalart (1988) DNS has been established as a research tool. It was shown, that the results of experiments and DNS are indistinguishable for a common setup, although the problem of realising boundary conditions in an experiment or a simulation remains (Moin & Mahesh, 1998). In recent years, experimental and numerical studies complement each other and both approach inspired research activities. Certain questions that arose from, but could not be answered within, experiments, were answered by numerical studies and vise versa. Each approach has its strength and unique possibilities. On one hand, the range of Reynolds numbers that can be explored in experiments is generally larger compared to the one in DNS. Similarly, a parameter study for a certain flow configuration can often be carried out at lower costs in experiments. On the other hand a DNS provides a detailed high-fidelity picture of the flow. In particular, all flow quantities are provided in high resolution in space and time. Another benefit of DNS is that modified equations or artificial boundary conditions can be realised to test a hypothesis or study certain aspects of turbulence (Jiménez & Moin, 1991; Jiménez & Pinelli, 1999; Flores & Jiménez, 2010).

To conclude, direct numerical simulation has been established as a research tool since many years (Moin & Mahesh, 1998). The computational costs and time requirements related to DNS are immense and limits its use to low or moderate Reynolds numbers. However, the simulations provide a wealth of data which would be difficult to obtain otherwise. In that sense, DNS complement experiments and are equally important to advance the understanding of turbulent flows.

2.2.3 Numerical method and validation

This thesis presents direct numerical simulations of flow over fixed and direct numerical simulations of flow with mobile particles. The simulations were carried out with the numerical code SUSPENSE developed by Markus Uhlmann. The numerical method is documented in several journal publications (Uhlmann, 2005*a*, 2008) as well as in a series of technical reports that provide additional detailed information on the numerical method (Uhlmann, 2003*a*,*b*, 2004, see also Uhlmann, 2006*b*).

The numerical method is based on a fractional-step method to numerically solve the incompressible Navier–Stokes equation. The temporal discretisation is semi-implicit, based on the Crank[1]–Nicolson[2] scheme for the viscous terms and a low-storage three-step Runge[3]–Kutta[4] procedure for the non-linear part (Verzicco & Orlandi, 1996). The spatial operators are evaluated by central finite-differences on a staggered grid. The temporal and spatial accuracy of this scheme is of second order.

A variant of the immersed boundary technique (Peskin, 1972, 2002) proposed by Uhlmann (2005*a*) is employed to discretise a wall roughened by spheres or mobile spheres moving freely in the flow. This method employs a direct forcing approach, where a localised volume force term is added to the momentum equations (2.3). The additional forcing term is explicitly computed at each Runge-Kutta sub-step as a function of the no-slip condition at the particle surface, without recurring to a feed-back procedure. The necessary interpolation of variable values from Eulerian grid positions to particle-related Lagrangian positions (and the inverse operation of spreading the computed force terms back to the Eulerian grid) are performed by means of the regularised delta function given by Roma, Peskin & Berger (1999).

All simulations carried out with SUSPENSE in this thesis employ a Cartesian grid with uniform isotropic mesh widths, i.e. $\Delta_x = \Delta_y = \Delta_z$. This ensures that the regularised delta function verifies important identities, such as the conservation of the total force and torque during interpolation and spreading (Roma *et al.*, 1999). For reasons of efficiency, forcing is only applied to the surface of the spheres, leaving the flow field inside the particles to develop freely.

In cases with freely moving particles the particle motion is determined by the Runge-Kutta discretised Newton equations of (2.9) to (2.10), which are explicitly coupled to the fluid equations. The hydrodynamic forces on a particle are readily obtained by integrating the additional volume forcing term over the particles boundary. Thereby, the exchange of momentum between the two phases cancels out identically and no spurious contributions are generated. The analogue procedure is applied for the computation of the hydrodynamic torque driving the angular particle motion.

During the course of a simulation, particles can approach each other closely. However, very thin inter-particle films cannot be resolved by a typical grid and therefore the correct build-up of repulsive pressure is not captured which in turn can lead to possible partial 'overlap' of the particle positions in the numerical computation. In the present study a contact model is applied to prevent such non-physical situations (cf. section §2.1.2.2).

[1]John Crank, English mathematical physicist, ⋆6 February 1916 †3 October 2006

[2]Phyllis Nicolson, English mathematician and physicist, ⋆21 September 1917 †6 October 1968

[3]Carl David Tolmé Runge, German mathematician, physicist, and spectroscopist, ⋆30 August 1856 †3 January 1927

[4]Martin Wilhelm Kutta, German mathematician, ⋆3 November 1867 †25 December 1944

The numerical methodology has undergone most extensive testing in order to guarantee high-quality results (Uhlmann, 2004, 2005*a*,*b*, 2006*b*, 2008; Doychev, 2010). Further credibility of the numerical method can be gained from the work of other researches which implemented the immersed boundary method in similar numerical frameworks (Lucci *et al.*, 2010, 2011; Lee & Balachandar, 2010). In particular Lee & Balachandar (2010) provide detailed validation studies on the numerical method and give information on the accuracy of drag and lift with respect to spatial resolution. The authors compare the results from the immersed boundary method with the results of a spectral element code. It might be noted, that the immersed boundary method employed to discretise the particles in the present study became popular in the past years, in particular it is described in a recent textbook on the subject (Loth, expected publication fall 2011).

In addition to the validations above, an effort was made to compare the result in this thesis to data from the literature wherever possible to further validate the present approach. The agreement is generally good and consistent with the literature.

Chapter 3

Open channel flow over fixed spheres

In this chapter turbulent open channel flow over an array of fixed spheres is studied. The results of two direct numerical simulations are analysed with special focus on the characterisation of force and torque on particles as well as the characterisation of flow structures related to force and torque. The setup of the simulations is described in §3.1. The flow field is analysed in §3.2 in terms of flow field statistics, spectra of the flow field and the three-dimensional time-averaged flow field. Sections 3.3 to 3.6 focus on the characteristics of force and torque and the characterisation of flow structures related to force and torque by discussing the statistics of particle force and torque, correlation functions of particle quantities in time and space-time and correlation functions between flow field and particle quantities. A summary, conclusions and recommendation for future work are given in §3.8.

Part of the results presented here have been published in conference contributions, internal reports and papers listed at the beginning of this thesis. In particular, the characterisation of force and torque on a particle was discussed in Chan-Braun *et al.* (2011).

3.1 Numerical setup

The flow configuration consists of turbulent open channel flow over a geometrically rough wall. The wall is formed by one layer of fixed spheres which are packed in a square arrangement (see figure 3.1). The distance between the particle centres is $D+2\Delta_x$, where D is the particle diameter and Δ_x is the grid spacing. At $y=0$ a rigid wall is located below the layer of spheres. As can be seen in figure 3.1 this rigid wall is roughened by spherical caps that can be defined as the part above $y=0$ of spheres located at $y=D/2-\sqrt{2}(D/2+\Delta_x)$, staggered in the streamwise and spanwise direction with respect to the layer of spheres above.

The physical and numerical parameters of the simulations are summarised in table 3.1. Note, that the values on the table differ marginally to those given in Chan-Braun *et al.* (2011) as they are based on somewhat longer time-series. The computational domain dimensions are $L_x/H \times L_y/H \times L_z/H = 12 \times 1 \times 3$, in streamwise, wall-normal and spanwise direction, respectively (cf. figures 3.2 and 3.3). An equidistant Cartesian grid with $3072 \times 256 \times 768$ grid points is employed.

One important parameter for the present flow configuration is the ratio between the domain height, H, and the spheres diameter, D. Ideally, a large H/D is desirable to ensure that the spheres can be considered as roughness and not as obstacles in a channel (Jiménez, 2004). However, from a practical point of view it is difficult to reach large values of H/D without increasing excessively the computational cost. Here, two cases are considered: case F10 with $H/D=18.3$ and a total of 9216 particles, and case F50 with $H/D=5.6$ and a total of 1024 particles above the wall. Periodic

Case	U_{bh}/u_τ	Re_b	Re_τ	D^+	D/Δ_x	Δ_x^+	N_p	$\tau_c U_{bH}/H$
F10	15.2	2870	188	10.8	14	0.77	9216	158
F50	12.3	2880	233	48.8	46	1.06	1024	126

Table 3.1: Setup parameters of simulations; U_{bH} is the bulk velocity based on the domain height H, U_{bh} is the bulk velocity based on the effective open channel height, h, defined as $h = H - 0.8D$, u_τ is the friction velocity, $\mathrm{Re}_b = U_{bH}H/\nu$ is the bulk Reynolds number with ν as the kinematic viscosity, $\mathrm{Re}_\tau = u_\tau h/\nu$ is the friction Reynolds number, $D^+ = Du_\tau/\nu$ is the particle diameter in viscous units, D/Δ_x is the resolution of a particle, Δ_x^+ is the grid spacing in viscous units, N_p is the total number of particles in a layer, τ_c is the time over which statistics were collected.

boundary conditions are applied in streamwise and spanwise directions. At the upper boundary a free-slip condition is employed. At the bottom boundary a no-slip condition is applied. The spheres are resolved using the immersed boundary method which is described in §2.2.3.

Figures 3.2 and 3.3 show the complexity of the flow in both cases. The flow field is visualised by iso-surfaces of the instantaneous streamwise velocity fluctuation. The common arrangement of low-speed (blue) and high-speed (red) streaks are observed, in case F10. Here, the spheres are small and the streaks differ little from those observed in flow over a smooth wall (cf. figure B.1 in appendix B). Although the size of the spheres in case F50 is about three times larger than in case F10, the overall picture is similar, except that in the figures small filaments of high and low speed appear to be preferentially aligned in between the particles and the streaks as well as there spanwise spacing seems somewhat smaller. This might be influenced by the choice of threshold or by the definition of the velocity fluctuation but could also stem from the increase roughness effect in case F50. The fact that large scale structures are damped as an effect of roughness has been previously reported for example by Flores & Jiménez (2006) and will be discussed in more detail in §3.2.2.

In order to scale the results, two quantities need to be specified: the friction velocity, u_τ, and the location of the virtual wall, y_0, since for a geometrically rough wall the position of the wall cannot be unambiguously defined (cf. Townsend, 1971; Raupach *et al.*, 1991). As discussed in detail in appendix C.1, it is chosen to define the position of the virtual wall as $y_0 = 0.8D$ throughout this study. The value of u_τ is defined by extrapolating the total shear stress $\tau_{tot} = \rho_f \nu \partial\langle u\rangle/\partial y - \rho_f\langle u'v'\rangle$ from above the roughness layer (where it varies linearly) down to the location of the virtual wall, y_0. Here, ν is the kinematic viscosity and ρ_f is the fluid density.

The effective flow depth, h, can be defined as the distance from the virtual wall to the top boundary, $h = H - y_0$. The bulk velocity based on the domain height, H, is defined as $U_{bH} = 1/H\int_0^H \langle u\rangle\,\mathrm{d}y$, the bulk velocity based on the effective flow depth is defined as $U_{bh} = 1/h\int_{y_0}^H \langle u\rangle\,\mathrm{d}y$. In the following angular brackets with sub-indexes, $\langle\cdot\rangle_{t,x,y,z,b,p}$, are used for averaging over time (t) along a certain coordinate direction (x, y, z) over periodically repeating boxes of geometry (b) and over the particle related data (p) respectively. Angular brackets without additional indexes, $\langle\cdot\rangle$, refer to quantities which are averaged over time as well as spatially over wall-parallel planes, i.e. along the x and z directions. The deviation of the instantaneous value of a quantity, ϕ, to the averaged value $\langle\phi\rangle$ is indicated by a prime as super index, e.g. ϕ'. When the root mean square value of a quantity is considered the sub-index rms is used, e.g. ϕ_{rms}.

The bulk Reynolds number, $\mathrm{Re}_b = U_{bH}H/\nu$, was kept constant at a value of 2870 and 2880 in cases F10 and F50, respectively. This corresponds to a friction Reynolds number, $\mathrm{Re}_\tau = u_\tau h/\nu \simeq 180$ in case of a smooth wall. In the present simulation the value for Re_τ increases to 188 in case F10 and to 235 in case F50. The grid resolution of the simulation was in both cases approximately equal

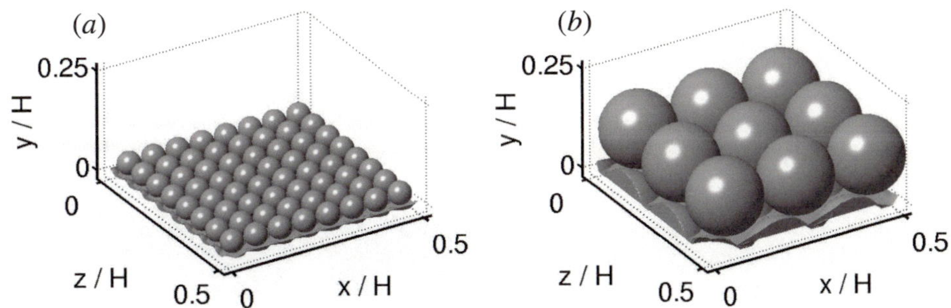

Figure 3.1: Close-up of a section of the computational domain with the geometry of the bottom wall consisting of a layer of fixed spheres arranged on a square lattice; (*a*) case F10; (*b*) case F50.

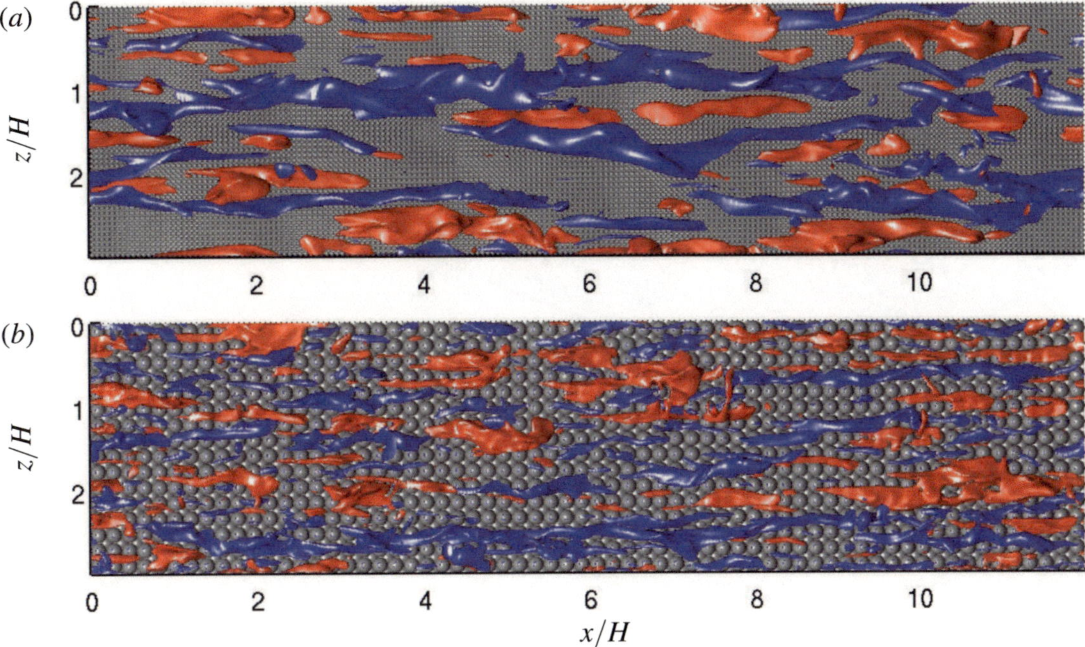

Figure 3.2: Top view of instantaneous flow field in case F10 (*a*) and case F50 (*b*). Red (blue) surfaces are iso-surfaces of the streamwise velocity fluctuation at values $+3u_\tau$ ($-3u_\tau$).

to the viscous length $\delta_\nu = \nu/u_\tau$ in all spatial directions. The resolution can therefore be qualified as exceptionally fine away from the wall and as reasonably fine in the vicinity of the rough wall. In the following, normalisation of a quantity, ϕ, with wall units will be denoted by a superscript $+$, e.g. ϕ^+.

The initial turbulent flow field of each simulation was taken from a similar simulation on a coarser grid. Subsequently, the simulation was run until the flow reached a statistically stationary state. The simulation was then continued for $\tau_c U_{bH}/H$ as shown in table 3.1 during which flow field statistics as well as particle data such as forces and torques were collected. Entire flow fields jointly with the particle data were stored at intervals of about H/U_{bH}. Based on these data a statistical analysis has been carried out. If not explicitly stated otherwise, the statistics shown for cases F10 and F50 stem from the data collected during the run-time of the simulation and are averaged over the entire domain

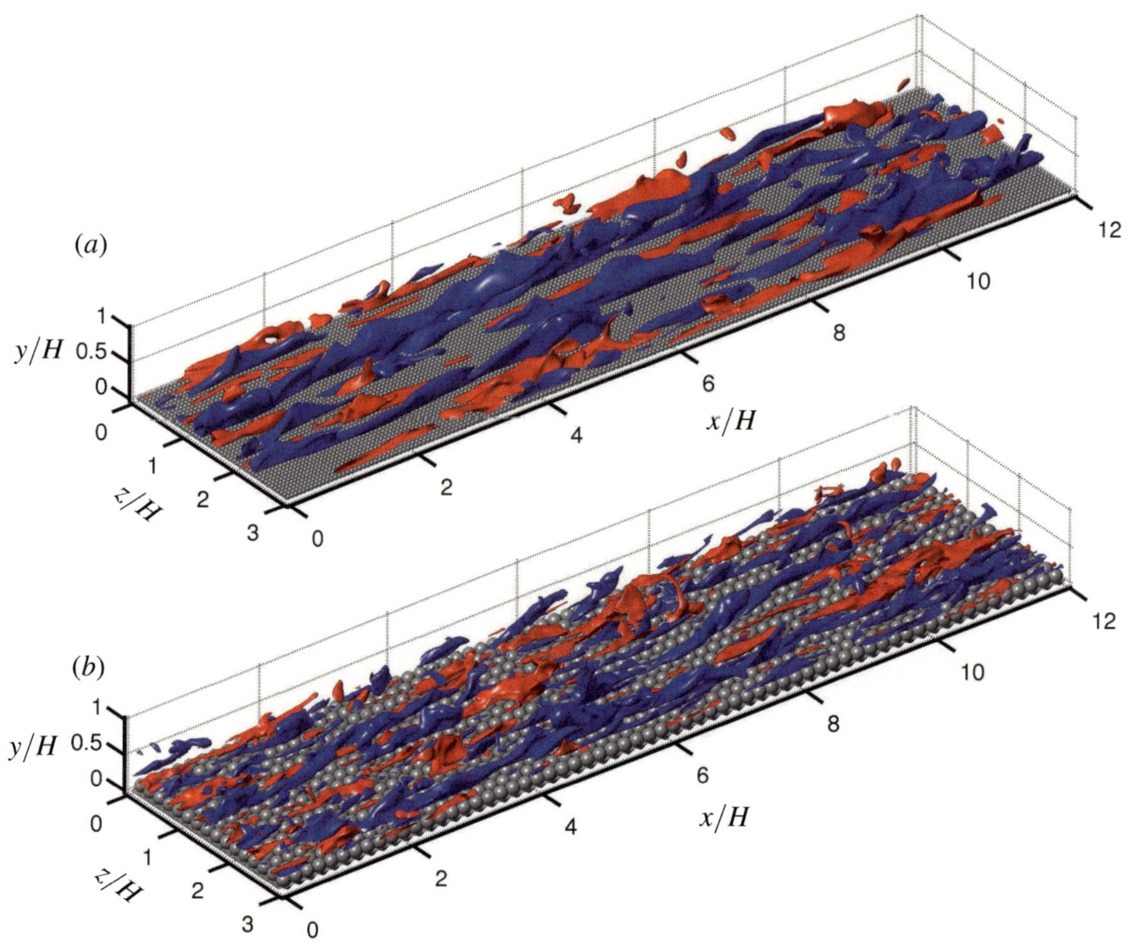

Figure 3.3: Instantaneous flow field in case F10 (*a*) and case F50 (*b*). Red (blue) surfaces are iso-surfaces of the streamwise velocity fluctuation at values $+3u_\tau$ ($-3u_\tau$).

including the region within the particles. Details on the different averaging procedures used and how they compare are provided in appendices C.2 and C.3.

3.2 Flow field characteristics

3.2.1 Flow field statistics

Figure 3.4 shows the profiles of the time and plane averaged streamwise velocity component, $\langle u \rangle$, as a function of the vertical coordinate. The results of case F10 and case F50 are compared with the reference case S180 of a smooth wall open channel flow at $\mathrm{Re}_b = 2880$ and $\mathrm{Re}_\tau = 183$, which has been recomputed for the present study (cf. appendix B for details). The profiles show the expected effect of roughness that is described in various textbooks (Schlichting, 1965; Pope, 2000). As the particle diameter D increases, while keeping the value of the bulk Reynolds number Re_b constant, the friction velocity increases; the streamwise velocity profile normalised by outer scales flattens (figure 3.4*a*), and the streamwise velocity profile normalised by viscous scales increasingly shifts towards lower values of $\langle u \rangle^+$ (figure 3.4*b*). Figure 3.4 shows that a logarithmic layer exists, if at all, only over a small range due to the low Reynolds number considered. The logarithmic law for the flow over a rough wall can be written as in the case of a smooth wall with an additional offset ΔU^+ that accounts for the roughness effect

$$\langle u\rangle^+ = \frac{1}{\kappa}\ln\left(\frac{y-y_0}{\delta_v}\right) + A - \Delta U^+ , \tag{3.1}$$

where the von Kármán[1] constant, κ, and A are constants obtained empirically to be $\kappa \approx 0.4$ and $A \approx 5.1$ (according to experimental findings summarised e.g. in Jiménez, 2004). From the profiles in figure 3.4 it appears that in case F10 the roughness effect is weak, while in case F50 a stronger roughness effect can be seen.

It is customary to quantify the roughness effect by using the equivalent sand grain roughness k_s (Schlichting, 1936). It can be obtained by a fit to the mean velocity profile in the logarithmic layer using the following equation

$$\langle u\rangle^+ = C\,\log_{10}\left(\frac{y-y_0}{k_s}\right) + B . \tag{3.2}$$

Here, the values of $B \approx 8.48$ and $C = 1/\kappa\,\ln(10) \approx 5.75$ are empirically obtained (cf. Shockling *et al.*, 2006). At high enough Reynolds numbers k_s becomes constant, i.e. $\lim_{\mathrm{Re}\to\infty} k_s = k_{s\infty}$, defining the so-called fully rough flow regime. The specific value of $k_{s\infty}$ is a quantity of the surface that depends on the characteristics of the roughness, such as shape, arrangement or roughness area ratio.

Flow over roughness can be classified as hydraulically smooth, transitionally rough or fully rough according to a small, moderate or high value of $k^+_{s\infty}$. Nikuradse (1933) gave values of $5 < k^+_{s\infty} < 70$ to define the transitionally rough flow regime. However, these values should be taken with care as the transition might be influenced by the specific characteristics of the roughness (cf. discussion in Bradshaw, 2000; Jiménez, 2004; Shockling *et al.*, 2006, among others). In particular, it has been speculated that a uniformly sized, structured arrangement of roughness elements as in the present case might lead to a sharp transition from hydraulically smooth to the fully rough regime (Colebrook, 1939; Jiménez, 2004).

Figure 3.5 shows the transition from the hydraulically smooth flow regime to the fully rough flow regime obtained in different experiments. It shows the offset ΔU^+ as a function of $k^+_{s\infty}$. At low values of $k^+_{s\infty}$ the effect of roughness should be negligible and correspondingly ΔU^+ approaches zero. In the fully rough flow regime the roughness effect should purely depend on $k^+_{s\infty}$. By comparison of equation (3.1) and (3.2), a formula for ΔU^+ can be derived for the fully rough regime,

$$\Delta U^+ = C\,\log_{10}\left(k^+_{s\infty}\right) - B + A , \tag{3.3}$$

with the constants A, B and C as above. The relation (3.3) above is shown in figure 3.5 jointly with results from experiments.

For the present simulations the values of ΔU^+ can be obtained by the vertical shift of the mean velocity profiles in the log-region of figure 3.4(*b*). They are 1.03 and 4.85 for cases F10 and F50, respectively. However, the value of $k_{s\infty}/D$ for the present arrangement of spheres is unknown. A table of values for $k_{s\infty}/D$ found in related studies on flow over spheres is provided in the appendix (cf. table C.1). Schlichting (1936), Ligrani & Moffat (1986) and Pimenta, Moffat & Kays (1975) found a value of $k_{s\infty}/D \sim 0.63$ for flow over spheres in hexagonal packing, while somewhat larger

[1]Theodore von Kármán (orig. Hungarian name: Szőllőskislaki Kármán Tódor), Hungarian–American aerospace engineer and physicist, ⋆ 11 May 1881 † 7 May 1963

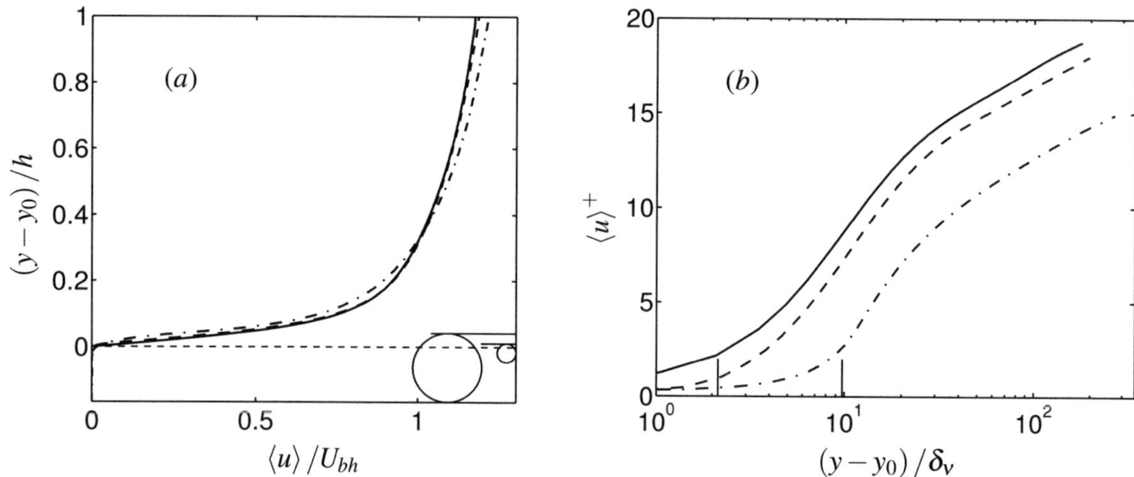

Figure 3.4: Time and spatially averaged streamwise velocity component, $\langle u \rangle$, of case F10 (– – –) and case F50 (– · –) in comparison with smooth wall open channel flow (———). (*a*) Normalised with U_{bh} as a function of $(y-y_0)/h$, (*b*) in semi-logarithmic scale normalised by δ_v and u_τ. The position of the particles top are marked with horizontal (*a*) and vertical (*b*) solid lines.

values were obtained by Singh *et al.* (2007) ($k_{s\infty}/D = 0.77$) and Detert *et al.* (2010*a*) ($k_{s\infty}/D = 0.81$). For flow over spheres in random packing, the obtained values vary in the range of $k_{s\infty}/D = 0.55$ to 0.85 (cf. Muñoz Goma & Gelhar, 1968; Grass *et al.*, 1991). Few studies exist that use values of $k_{s\infty}/D = 1$ for structured (Einstein & El-Samni, 1949) or random arrangements (Nakagawa & Nezu, 1977). Figure 3.5 shows the pair of values $(k_{s\infty}^+, \Delta U^+)$ for cases F10 and F50 when approximating $k_{s\infty}/D$ by the value found for a hexagonal packing, i.e. $k_{s\infty}/D = 0.63$. The error-bars indicate the range of $k_{s\infty}/D = 0.55$ to 1 as found in the literature. It can be seen that case F10 approaches the hydraulically smooth flow regime while case F50 is in the transitionally rough flow regime.

A similar conclusion can be reached by analysing the profiles of the root-mean-square of the velocity fluctuations normalised with u_τ that are shown in figure 3.6(*a*). In case F10, the profiles of the three velocity components almost collapse with the smooth wall results, indicating that, indeed, the flow over the relatively small roughness elements in this case can be considered as nearly hydraulically smooth. In case F50, some differences with respect to the smooth wall case are evident. The near-wall peak in the streamwise fluctuation profile decreases but it is still visible. This indicates that the flow is in the transitionally rough flow regime, since experiments in the fully rough flow regime present no clear peak (see for example figure 5 of Jiménez, 2004). The wall-normal and spanwise fluctuations present slightly higher values near the wall than the corresponding ones in the smooth wall case. Therefore, the anisotropy of the fluctuations near the wall is smaller than in the smooth wall case. This tendency of roughness to make the fluctuations more isotropic is a phenomenon which has been often reported in the literature (e.g. Poggy, Porporato & Ridolfi, 2003; Orlandi & Leonardi, 2008). Also in case F50, above $(y-y_0)/h \sim 0.4$ all three components agree well with the values of the smooth wall case.

Orlandi & Leonardi (2008) discuss the velocity shift ΔU^+ as a function of the root mean square value of the wall-normal velocity fluctuations, v_{rms}, at the roughness crest. The present simulations result in pairs $(v_{\mathrm{rms}}, \Delta U^+)$ of $(0.10, 1.03)$ $(0.46, 4.85)$ for case F10 and F50 respectively which agree well within the scatter of the reported data (graph not shown).

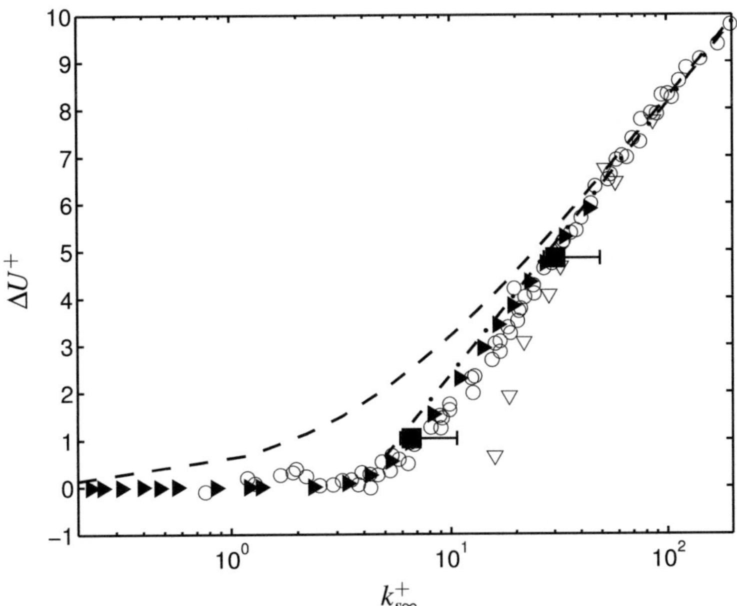

Figure 3.5: Roughness function for several transitionally rough surfaces as a function of the Reynolds number $k^+_{s\infty}$, adapted from Jiménez (2004). Symbols and lines correspond to Nikuradse (1933), uniform sand, pipe flow (○), Ligrani & Moffat (1986), uniform densely-packed spheres, boundary layer (▽), Shockling *et al.* (2006), honed aluminium, pipe flow (▶), present simulation with $k_{s\infty}/D = 0.63$ (■) error-bars show the range of $k_{s\infty}/D = 0.55$ to 1, relation $\Delta U^+ = 5.75 \log(1 + 0.26 k^+_{s\infty})$ as proposed by Colebrook (1939) (– – –), relation (3.3) with $A = 5.1, B = 8.48$ and $C = 5.75$ (– · –).

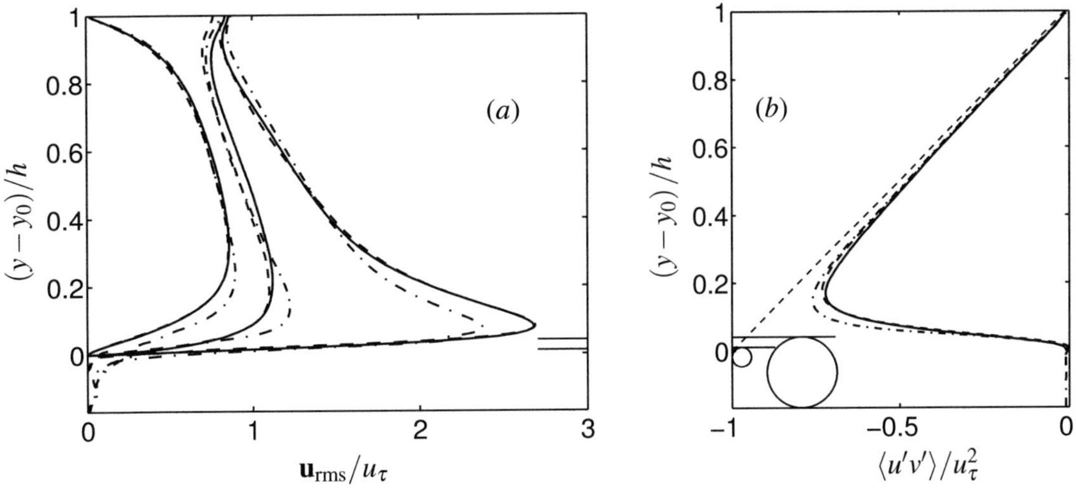

Figure 3.6: (*a*) Root-mean-square of velocity fluctuations of case F10 and case F50 normalised by u_τ in comparison with results of smooth wall as a function of wall distance. Curves from left to right: wall-normal (v_{rms}/u_τ), spanwise (w_{rms}/u_τ) and streamwise (u_{rms}/u_τ). (*b*) Distribution of Reynolds shear stress $\langle u'v' \rangle$ as a function of wall distance. Legend as in figure 3.4. The position of the particle tops are marked with horizontal solid lines. In (*b*) the straight dashed line is included to guide the eye.

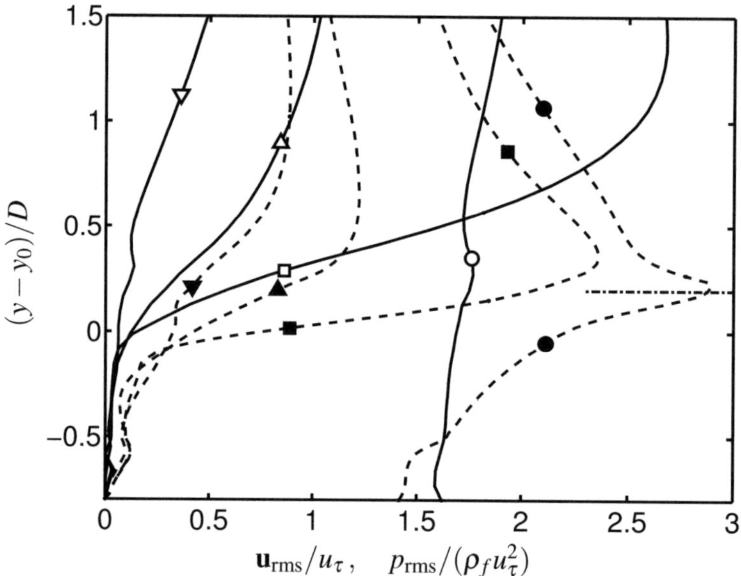

Figure 3.7: Root-mean-square of velocity and pressure fluctuations of case F10 and case F50, normalised by u_τ and $\rho_f u_\tau^2$ respectively, as a function of $(y-y_0)/D$. Averaging has been carried out over fluid cells only, for details see appendix §C.2. Lines with symbols indicate u_{rms}/u_τ (□), v_{rms}/u_τ (▽), w_{rms}/u_τ (△), and $p_{\text{rms}}/\rho_f u_\tau^2$ (○). Solid lines and empty symbols correspond to case F10, dashed lines and full symbols correspond to case F50, horizontal line indicates the position of particle tops in both cases.

Figure 3.6(*b*) shows the profiles of the Reynolds stress, $\langle u'v' \rangle$, normalised by u_τ^2. The Reynolds stress profile $\langle u'v' \rangle$ of case F10 nearly collapses with the profile of the smooth wall case. In case F50 a slight increase and a small shift towards the wall of the near-wall peak can be seen which could be an effect of the higher value of Re_τ in this case.

In order to study the near wall behaviour of the velocity fluctuations, a close-up of the profiles shown in figure 3.6(*a*), is plotted as a function of $(y-y_0)/D$ in figure 3.7. Additionally, the profiles of the root-mean-square of the pressure fluctuations, $p_{\text{rms}}/(\rho_f u_\tau^2)$, are included. Note, that in contrast to figure 3.6 the profiles shown in figure 3.7 are obtained from snapshots of the flow field and obtained by averaging over cells outside of the particles as described in detail in §C.2 and §C.3. The amplitudes of the fluctuations of the three velocity components present similar values below the virtual wall, $(y-y_0)/D < 0$. These are much smaller than the values above the virtual wall, and they are somewhat larger in case F50 ($\mathbf{u}_{\text{rms}} \sim 0.1 u_\tau$) compared to F10 ($\mathbf{u}_{\text{rms}} < 0.05 u_\tau$). On the contrary, the pressure fluctuations within the roughness layer for both cases present values which are similar to the values above the roughness layer.

Recall that also in the case of a smooth wall the pressure fluctuations are non-zero at the wall (Kim, Moin & Moser, 1987; Kim, 1989). Near the top of the roughness elements, i.e. around $(y-y_0)/D = 0.2$, the profiles of p_{rms} exhibit a peak which is barely visible in case F10 and more pronounced in case F50. Here, the value of the peak is higher by a factor of two compared to case F10. Note, that the peak is, to some extent, a consequence of the three-dimensionality of the time-averaged flow field around the particle; this point is further elaborated in §3.2.3. The pressure fluctuation profiles of case F10 and F50 (when plotted as a function of $(y-y_0)/h$) approach each other with increasing wall distance (not shown). They converge to the profile obtained in the case of a smooth wall in the outer part of the flow.

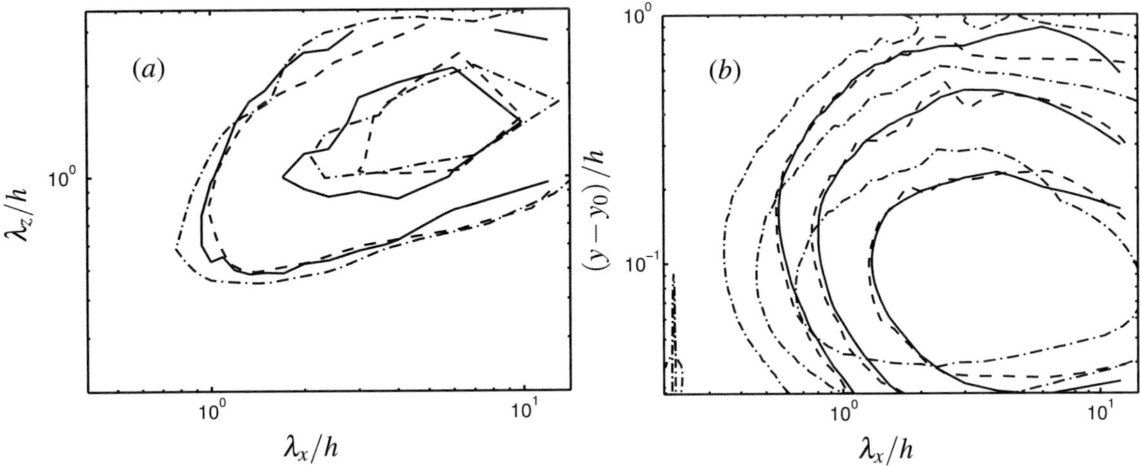

Figure 3.8: Pre-multiplied spectra of the streamwise velocity normalised by outer flow scales. (*a*) As a function of streamwise and spanwise wave length at $(y-y_0)/h = 0.5$. Lines show the iso-contours at 1/3 and 2/3 of the maximum $|\kappa_x \kappa_z \widehat{u}\widehat{u}^*|_{\max} h^2/U_{bh}^2$ in the smooth wall reference case. Logarithmic smoothing was applied to the contours of case F10 and case F50. (*b*) As a function of streamwise wave length and wall distance, the spectra are integrated over the spanwise wavelengths and multiplied by the streamwise wave number. Lines show iso-contours at $1/8$, $1/4$ and $1/2$ of the maximum $|\kappa_x \int \widehat{u}\widehat{u}^* \mathrm{d}\kappa_z L_z|_{\max} h/U_{bh}^2$ in the smooth wall reference case. In both panels the lines show smooth wall reference case (———), case F10 (– – –) and case F50 (– · –). The figure is analogous to figure 6 in Flores & Jiménez (2006).

3.2.2 Flow field spectra

The following discussion focuses on the streamwise velocity spectra in case F10 and case F50 in comparison with a smooth wall reference case. The two-dimensional streamwise velocity spectrum as a function of time and wall-normal distance is defined as $\widehat{u}\widehat{u}^*$, where $\widehat{u}(\kappa_x, y, \kappa_z, t)$ is the streamwise velocity component, Fourier transformed in wall-parallel planes, $\widehat{u}^*(\kappa_x, y, \kappa_z, t)$ its complex conjugate, and κ_x and κ_z the wave number in x and z direction, respectively. The spectra in case F10 and case F50 are ensemble averaged over the number of collected flow fields (cf. §C.3). The smooth wall reference stems from a simulation employing a pseudo-spectral method that solves the Navier–Stokes equations based on the numerical algorithm of Kim *et al.* (1987), i.e. based on Fourier expansion (with dealiasing) in the periodic streamwise and spanwise directions, and Chebyshev polynomials in wall-normal direction (see appendix B, PS-S180, for details).

The two-dimensional pre-multiplied streamwise velocity spectra, $\kappa_x \kappa_z \widehat{u}\widehat{u}^* h^2/U_{bh}^2$, of case F10 and case F50 and the smooth wall case at $(y-y_0)/h = 0.5$ is shown in figure 3.8(*a*). The lines show the iso-contours at 1/3 and 2/3 of the maximum of the spectra $|\kappa_x \kappa_z \widehat{u}\widehat{u}^*|_{\max} h^2/U_{bh}^2$ in the smooth wall case. In case F10 and case F50, logarithmic smoothing was applied to the spectra to reduce the noise caused by the limited number of samples. The statistical convergence does not allow to infer about details of the spectra, however it appears that the iso-contours of all three cases agree within the statistical uncertainty. This observation is in line with the results of Flores & Jiménez (2006) who obtained an overlap of contour lines in an analogous plot for simulations that are similar in their friction Reynolds numbers but differ in the roughness effect related to velocity disturbances imposed at the wall.

The maxima of the spectra in figure 3.8(*a*) are located in the upper range of wavelengths. The iso-contour at 2/3 of the maximum in the smooth wall case are of closed shape. In contrast, the iso-contours at 1/3 of the maximum are not fully contained in the given range of streamwise lengths scales.

This indicates, that a non-negligible amount of energy is contained in streamwise length scales that fill the computational domain in streamwise direction. Thus the present box size, which is comparable to the one of Kim *et al.* (1987), is small with respect to the largest energy containing scales of motion. This result is known from the literature (see for example discussion in del Álamo & Jiménez, 2003 and figure 6 of Flores & Jiménez, 2006). Larger domain sizes would be preferable, however, this comes with an increase in computational costs. The present domain size is a compromise between computational costs and bias due to the restrictions of the largest scales. Comparison to experimental evidence is generally found to show that the domain size is large enough for most statistics considered in this study, possible bias due to the limitations of the largest flow structures on the results will be pointed out in the respective sections below.

The pre-multiplied streamwise velocity spectra, $\kappa_x \int \widehat{u}\widehat{u}^* \mathrm{d}\kappa_z L_z h/U_{bh}^2$, of the smooth wall case, case F10 and case F50 are shown in figure 3.8(b) as a function of streamwise wavelength and wall-normal distance, $(y-y_0)/h$. As a result of the integration over the spanwise wave numbers, the convergence of the spectra in figure 3.8(b) is better compared to those shown in figure 3.8(a). The lines show iso-contours at 1/8, 1/4 and 1/2 of the maximum $|\kappa_x \int \widehat{u}\widehat{u}^* \mathrm{d}\kappa_z L_z|_{\max} h/U_{bh}^2$ in the smooth wall reference case. The agreement of the contour lines in case F10 to those of the smooth wall reference case is good, especially for smaller wave lengths. In contrast to this, the contours of case F50 deviate from the results of the smooth wall and case F10. Here, a shift of the maximum towards smaller wave lengths is observed. Additionally, the spectrum in case F50 is damped for large wave length in the region $(y-y_0)/h < 0.2$. The shift to smaller wave lengths might be explained by a friction Reynolds number effect. That is, a collapse of the profiles might be expected when the pre-multiplied spectra is plotted for inner scales, e.g. as in figure 1(a) del Álamo & Jiménez (2003). Here, the spectra is plotted in outer scales to study the scaling of the large scales structures, thus with an increase in the Reynolds number a shift of energy towards smaller scales as well as a shift of the near wall peak closer to the wall might be expected. The damping of the largest flow structures close to the rough wall is in agreement with the observation of Flores & Jiménez (2006) which was linked to the effect of roughness on the large scales.

Figure 3.8(b) reveals a peak in the contours of case F50 at $\lambda_x/h = 0.22$ which coincides with the spacing of the particles ($0.22h$) in that case. This indicates, that for $(y-y_0)/h < 0.1$, flow scales with streamwise wavelength that equal the particle spacing, contribute noticeably to u_{rms}. A part of this contribution is probably due to the three-dimensional time-averaged flow field which will be discussed in the section below.

3.2.3 Three-dimensional time-averaged flow field distribution

Since the geometry of the roughness is three-dimensional, the time-averaged flow field in the near wall region also varies in all three directions. In the following some characteristics of the time-averaged flow field obtained from the flow fields collected during run-time of case F10 and case F50 are discussed. In addition to the averaging in time the fields were averaged over periodically repeating boxes centred on the particles (henceforth indicated by the symbol $\langle\cdot\rangle_{tb}$). For simplicity this will be simply referred to as time averaging below. Figure 3.9 shows the distribution of the three-dimensional time-averaged streamwise velocity, $\langle u \rangle_{tb}$, for both cases. Two different (x,y)-planes are shown. The first one contains the centre of the particles of one streamwise row (figures 3.9a,c). The second one is located in between two streamwise rows of particles (figures 3.9b,d). As can be expected the flow field is very different from a single sphere in an unbounded turbulent flow or in a channel close to a wall (Bagchi & Balachandar, 2004; Zeng *et al.*, 2008).

The sheltering effect of the neighbouring particles causes the flow velocity to decrease rapidly close to the roughness tops and leads to marginal flow velocities within the roughness layer. The highest velocity gradients are produced in the vicinity of the roughness tops. Similar observations were made in the experiments of Pokrajac & Manes (2009) who studied a comparable particle arrangement at bulk Reynolds numbers of order 10^4 and a ratio of $h/D \approx 3.5$. Also similar to their results is the formation of a re-circulation between two spanwise rows of spheres that extends over the entire spanwise direction. The shape of the re-circulation is similar in both of our present cases, however the strength differs. In case F50, the backflow velocities reach values as low as $\langle u \rangle_{tb} \approx -0.4u_\tau$ in figure 3.9(c) and $\langle u \rangle_{tb} \approx -0.2u_\tau$ in figure 3.9(d). In case F10 the magnitude of the backflow velocity is below $0.05u_\tau$. The re-circulation can also be observed in figure 3.10 that shows streamlines of the mean flow projected into the same planes shown in figure 3.9, i.e. by computing the streamlines using only $\langle u \rangle_{tb}$ and $\langle v \rangle_{tb}$, together with contours of the time-averaged pressure field. The pressure distributions in both simulations are similar. However, in case F10 the magnitude of the pressure, $\langle p \rangle_{tb} / \left(\rho_f u_\tau^2\right)$, is a factor of two smaller compared to case F50. Please note, that the three-dimensional time-averaged flow is not fully converged and at the location of the planes shown in figure 3.10 there is a weak net flow in the spanwise direction with a maximum amplitude of $\langle w \rangle_{tb} \approx 4 \cdot 10^{-4} U_{bh}$ ($6 \cdot 10^{-4} U_{bh}$) in case F10 (F50). This net flow is within the range of the statistical uncertainty.

A question of interest is how far the three-dimensionality of the bottom wall directly influences the flow. Figure 3.10 already shows that one particle diameter above the roughness tops the time-averaged pressure field is still visibly affected. In order to quantify the effect of three-dimensionality, the difference between the time-average of a field $\langle \phi \rangle_{tb}$ (where ϕ can stand for either pressure or one of the velocity components) and its time and plane-averaged value, $\langle \phi \rangle$, can be defined, viz.

$$\phi'' = \langle \phi \rangle_{tb} - \langle \phi \rangle \,. \tag{3.4}$$

Note, that the quantity, ϕ'', as defined by (3.4) is sometimes called spatial disturbance in the context of the double-averaging methodology (Nikora *et al.*, 2001). The corresponding root-mean-square value of ϕ'' can be computed using equation (3.4) as $\phi''_{\mathrm{rms}} = \sqrt{\langle \phi'' \phi'' \rangle}$. In both cases, F10 and F50, the standard deviation of pressure p''_{rms} drops by several orders of magnitude in between $y = D$ and $y = 2D$, as shown in figure 3.11. The same is true for the velocity field (not shown). Therefore, the time-averaged flow statistics appear to be essentially one dimensional beyond wall distances of $2D$. This is somewhat smaller than the values reported in previous investigations of flow over rough walls (cf. Jiménez, 2004) which might be related to the low values of D^+ and Re_τ considered.

3.3 Statistics of particle force

The hydrodynamic force, $\mathbf{F}$, acting on a particle is defined as

$$\mathbf{F} \;=\; \int_\Gamma \boldsymbol{\tau} \cdot \mathbf{n} \,\mathrm{d}\Gamma - \int_\Gamma p^{tot} \mathbf{n} \,\mathrm{d}\Gamma, \tag{3.5}$$

where Γ is the sphere's surface, $\mathbf{n}$ is the surface normal vector, $\boldsymbol{\tau} = \nu \rho_f \left(\partial_j u_i + \partial_i u_j\right)$ is the viscous stress tensor and p^{tot} is the pressure. The latter can be split into two parts $p^{tot} = p + p_l$, where p_l represents the linear variation in streamwise direction which results from the imposed pressure-gradient that drives the flow, and p corresponds to the three-dimensional instantaneous fluctuation. The first term on the right hand side of equation (3.5) is the force due to viscous stresses, the second term is the force due to pressure. A sketch that illustrates the definition of the force on a particle can be seen in figure 3.12(a).

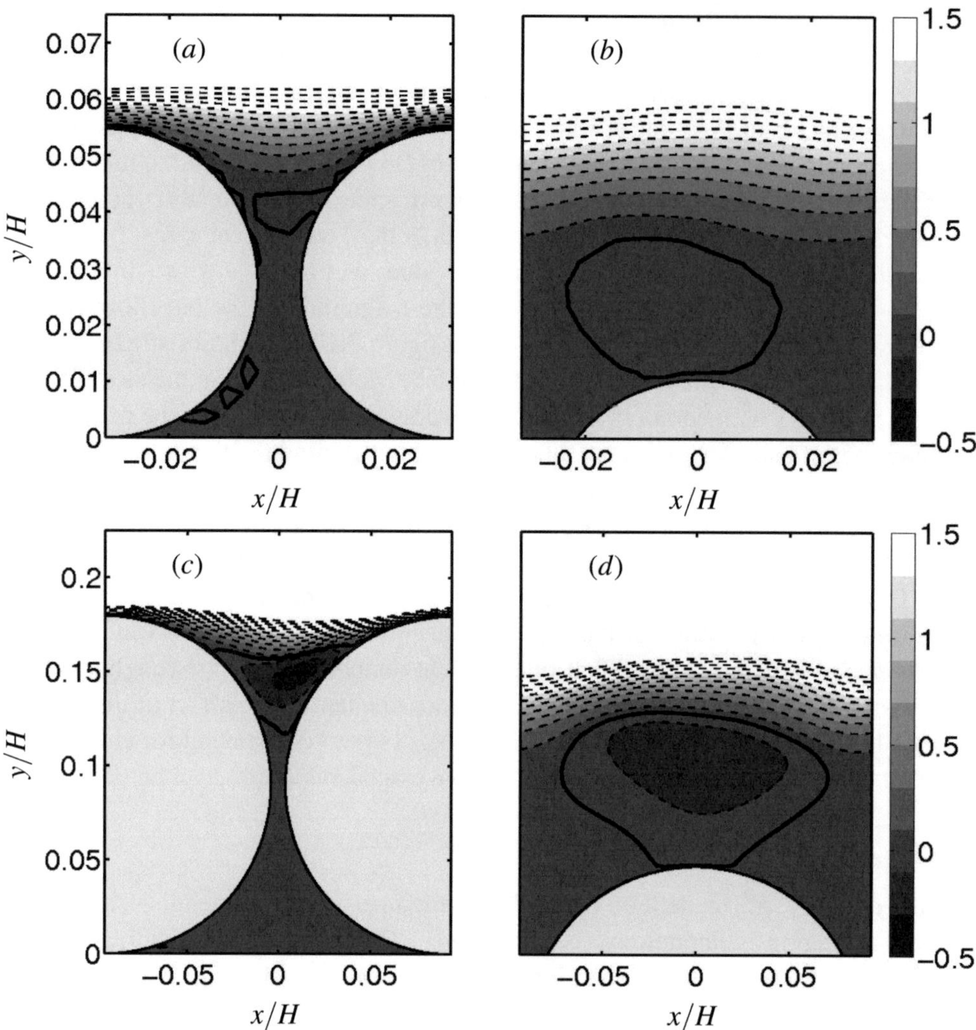

Figure 3.9: Distribution of the time-averaged streamwise velocity in a periodic cell, $\langle u \rangle_{tb}$, normalised by u_τ in *x*-*y* planes for case F10 (*a*,*b*) and case F50 (*c*,*d*). (*a*,*c*) Plane through particle centres, (*b*,*d*) plane centred between spheres. Lines show iso-contours of $\langle u \rangle_{tb}$ at values of -0.5 to 2.1 in steps of 0.2 (– – –), streamwise velocity contour at -10^{-3} (———). The direction of the bulk velocity is from left to right in all panels.

In order to scale the hydrodynamic forces, reference quantities need to be defined. For the present case of particles within a roughness layer the subject is a matter of discussion and several definitions have been proposed in the literature (see Hofland *et al.*, 2005). Here, the reference force is defined as $F_R = \rho_f u_\tau^2 A_R$ with the reference area $A_R = L_x L_z / N_p$.

Table 3.2 summarises the particle force statistics of the two cases, where C_{Fi} is the mean force on a particle in x_i-direction normalised by F_R. As can be seen, the mean values of the forces acting in the streamwise direction (henceforth also called "drag") and the wall-normal direction ("lift") are positive. Since the mean forces are directly related to the mean flow through the time-averaged version of equation (3.5), it is possible to shed some light onto the mechanisms that lead to mean drag and lift by analysing figures 3.9 and 3.10. A more detailed picture can be obtained from figure 3.13 and

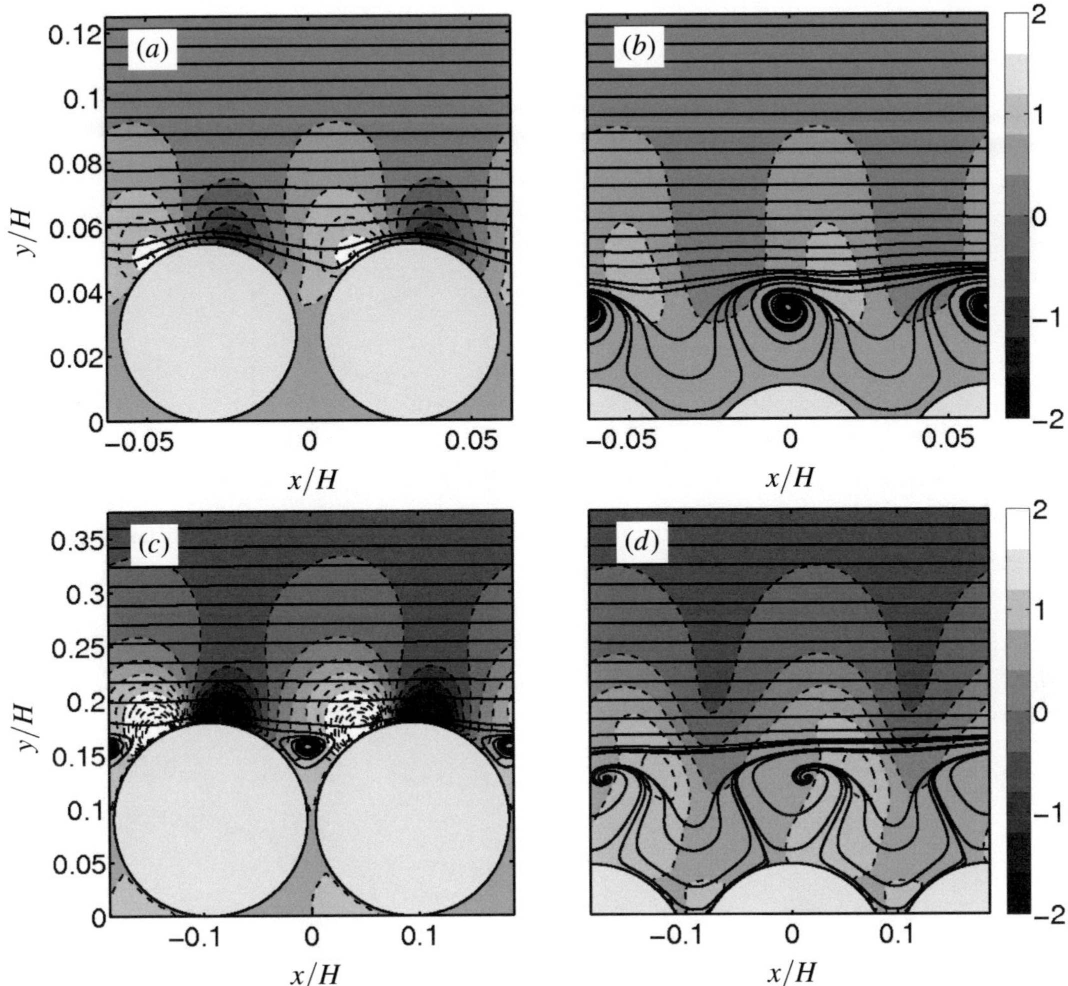

Figure 3.10: As figure 3.9 but showing the time-averaged pressure field and corresponding streamlines. Lines shows iso-contour lines of pressure $\langle p\rangle_{tb}/\left(\rho_f u_\tau^2\right)$ from values of -4 to 4 in steps of 0.4 (– – –) and streamlines in the plane computed from $\langle u\rangle_{tb}$ and $\langle v\rangle_{tb}$ (———). The direction of the bulk velocity is from left to right in all panels.

figure 3.14 which show the distribution on the sphere's surface of the stress leading to drag, τ_D, and lift, τ_L, viz.

$$\tau_D = \left(\langle\boldsymbol{\tau}\rangle_t\cdot\mathbf{n} - \langle p^{tot}\rangle_t\mathbf{n}\right)\cdot\mathbf{e}_x\,, \tag{3.6}$$

$$\tau_L = \left(\langle\boldsymbol{\tau}\rangle_t\cdot\mathbf{n} - \langle p^{tot}\rangle_t\mathbf{n}\right)\cdot\mathbf{e}_y\,, \tag{3.7}$$

where $\mathbf{e}_i$ is the unit vector in the x_i-direction. The stresses in figures 3.13 and 3.14 are normalised by F_R/A_{sph}, where $A_{sph} = \pi D^2$ is the surface area of the sphere. By virtue of this normalisation the total integral of the quantities shown in the figures yields the force coefficients C_{Fi} given in table 3.2. Please note that the results of case F50 appear less smooth due to the smaller number of particles, and therefore a smaller number of samples.

Figure 3.13 shows that the local stress contributing to drag is similarly distributed over the particle surface in both of the present flow cases F10 and F50. One can observe a region of strong positive

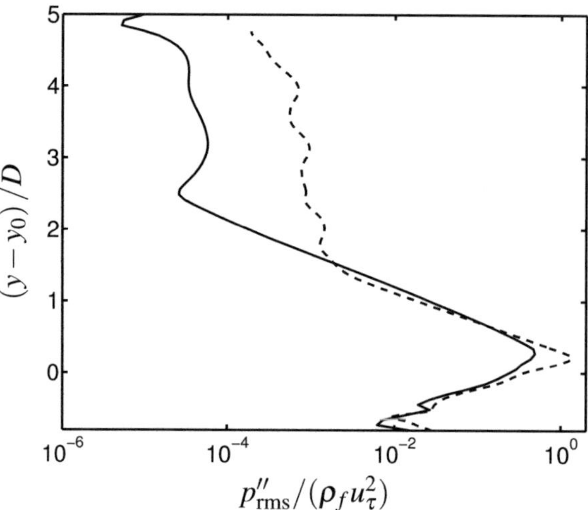

Figure 3.11: Three-dimensionality of the time-averaged flow field as a function of wall distance quantified via p''_{rms} (defined in equation 3.4) in case F10 (———) and case F50 (– – –).

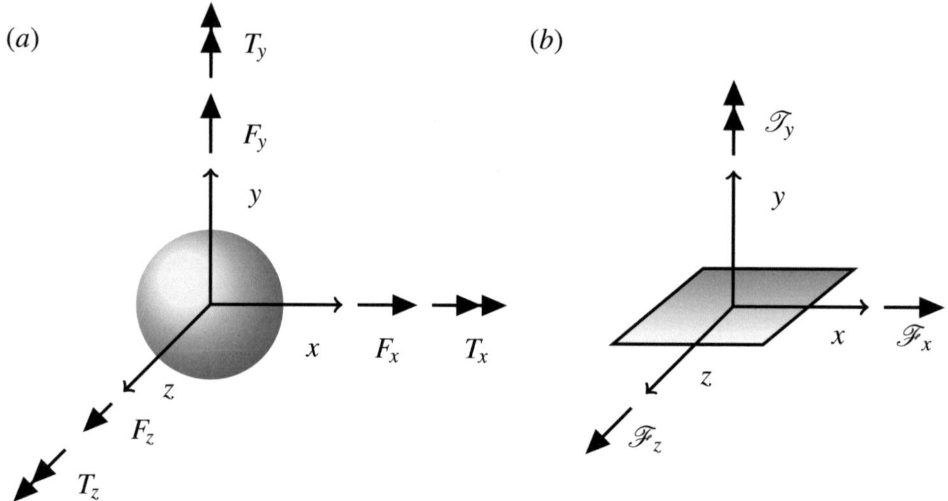

Figure 3.12: Sketch illustrating the definition of force (—►) and torque (—►►) on a particle (*a*) and a square surface element in a smooth wall channel (*b*)

values with the largest magnitude centred around a position slightly upstream of the particle tops. From figure 3.13(*a*,*c*) it can be seen, that this region of high positive local contributions to drag is slightly elongated in the spanwise direction. It results from the wall-normal gradients of the average streamwise velocity component, which are particularly important in the upper part of the sphere, as well as from the high pressure values found near the upstream side of each sphere (cf. figures 3.9 and 3.10). On the downstream side of the particles, still in the upper hemisphere, a smaller region with weak negative contributions to drag is found, as a result of the re-circulation region. In most of the lower (near-wall) half of the spheres, the contour lines of the local drag contribution are roughly oriented in the wall-normal direction, changing sign slightly downstream of the cross-stream plane passing through the particle centre. In this context it should be noted, that the driving

Case	C_{Fx}	C_{Fy}	C_{Fz}	α_F	σ_{Fx}/F_R	σ_{Fy}/F_R	σ_{Fz}/F_R	S_{Fx}	S_{Fy}	S_{Fz}	K_{Fx}	K_{Fy}	K_{Fz}
F10	1.04	0.19	0.00	11°	0.57	0.20	0.66	0.18	1.80	0.01	10.13	19.08	9.92
F50	1.15	0.37	0.00	18°	1.32	0.66	1.26	0.06	0.26	0.01	4.98	5.68	4.29

Table 3.2: Statistics of force on particles in case F10 and case F50, where $C_{Fi} = \langle F_i/F_R \rangle$ is the normalised mean force component in the x_i-direction, $\alpha = \arctan(C_{Fy}/C_{Fx})$ is the angle of the resulting force with respect to the x-axis, σ_{Fi} is the standard deviation of the force in x_i, S_{Fi} and K_{Fi} are the skewness and kurtosis of the respective force component.

pressure gradient $\mathrm{d}p_l/\mathrm{d}x < 0$ makes a weak but non-negligible contribution to the drag which can be quantified as approximately 2% (9%) of C_{Fx} in case F10 (F50). Therefore, non-negligible values of local contributions to drag are expected even in relatively quiescent regions, as is the case inside the roughness layer.

The qualitative and quantitative similarity of the distribution of τ_D in both cases F10 and F50 results in similar values for the drag coefficient in both cases (cf. table 3.2). In particular, the overall drag coefficient C_{Fx} in case F10 is close to unity, increasing to 1.15 in case F50. These values are a result of the weak contribution of the drag on the rigid wall below the layer of spheres to the total drag on the wall and the choice of the reference force. The drag coefficient as defined in the present study can be approximated as

$$C_{Fx} \approx \frac{V_{fb} + V_p}{hA_R}, \tag{3.8}$$

where V_{fb} is the total volume occupied by fluid in a periodic box around a particle, V_p is the volume occupied by a particle. The approximation (3.8) neglects the streamwise component of the shear force acting on the bottom wall in addition to the drag due to the periodic part of the pressure acting on the spherical caps. Evaluating this geometrical relation (3.8) yields 1.04 (1.15) for case F10 (F50).

Positive values for the lift coefficient, as observed in the present simulations (cf. table 3.2), can be explained by two mechanisms. The approaching flow accelerates in the frontal part until the top of the sphere and from then on it decelerates. This fact is reflected in the curvature of the streamlines (figures 3.9*a*,*c*), yielding a pressure distribution which exhibits lower values of pressure near the particles tops (figures 3.10*a*,*c*), and therefore a positive lift. In addition to pressure, shear might lead to a positive lift. As can be seen in figure 3.9, the flow field above the spheres is asymmetric (with respect to a cross-sectional plane through the particle centres) as a result of the re-circulation behind the particles. Therefore, the friction on the upstream side of the particle (in the upper hemisphere) is expected to be higher compared to the corresponding friction on the downstream side, contributing positively to the lift on the particle. The pressure differences as well as the asymmetry of the flow seem to be more pronounced in case F50 than in case F10 and might explain the observed increase in the lift coefficient.

Figure 3.14 shows the distribution on the sphere's surface of the stress leading to lift, τ_L. The shape of the contours is again similar in both cases, however, the magnitude of the stress τ_L seems to be significantly larger in case F50, leading after integration to the factor of two presented in table 3.2. The spatial distribution is characterised by one dominant patch of each, positive and negative, values of τ_L, the maximum of both being located on the (x,y) symmetry plane, the former (positive) near the particle top, the latter (negative) shifted upstream by approximately 45° (30°) in case F10 (F50). From the contours, it appears that the flow below the virtual wall contributes little to the lift.

In order to quantify the contribution integrated from the bottom of a particle up to a certain fraction of its diameter, we can define a cumulative function

$$\mathscr{S}_\phi(y) = \frac{D}{2} \int_0^y \int_0^{2\pi} \tau_\phi(y,\theta)\, \mathrm{d}\theta \mathrm{d}y\,, \tag{3.9}$$

where $\tau_\phi(y,\theta)$ stands for either τ_L or τ_D evaluated at a position on the sphere's surface given by the wall-distance y and an azimuthal angle θ in the wall-parallel plane. Figure 3.15 shows $\mathscr{S}_D$ and $\mathscr{S}_L$ normalised by the net values of lift and drag, respectively. The contribution to the net drag by the flow in the lower half of the sphere is small in both cases, the cumulative drag value increasing monotonically and with increasing slope from the wall to the top of the sphere. Conversely, the cumulative contribution to the lift first increases with increasing wall-distance up to values of approximately 25% (40%) of the total in case F10 (F50) at $y \approx 0.5D$, before decreasing again to a small value at $y \approx 0.9D$. Beyond that, in a small area surrounding the top of the sphere, is where most of the net lift is generated. In case F50, the lift increases with respect to case F10 more than the drag, which leads to a higher angle, $\alpha = \arctan(C_{Fy}/C_{Fx})$, of the resulting force (cf. table 3.2).

The spanwise force should be zero for symmetry reasons. In both cases the calculated mean spanwise force coefficient is more than two orders of magnitude lower than the drag coefficient. This fact provides confidence in the convergence of the statistics.

Additional support to the mean forces just discussed is provided by comparison to experimental measurements performed in a somewhat similar configuration by Hall (1988). In that study, the mean lift on a particle near a boundary was measured in a wind tunnel with smooth as well as rough walls. The interesting case for the present discussion consisted of a sphere of diameter D placed in between spanwise rods of diameter D_r evenly spaced out with a distance D_r. Figure 3.16 presents the comparison of the mean lift normalised by $\rho_f \nu^2$ as a function of D^+, between the values obtained in the experiments of Hall (1988) and the present simulations. In spite of the different setups, the lift obtained in case F10 is perfectly consistent with the measurements while the lift obtained in case F50 is somewhat lower. The reason for this might be that in the setup of the simulations the neighbouring spheres are closer producing an increased sheltering effect. This is also supported by the experimental observation that lower lift values are obtained when the value of D_r/D is increased (Hall, 1988).

In contrast to the direct relation between the mean flow field and the mean forces on a particle, a similar straightforward relation between the statistics of the fluid velocity fluctuations and of the particle force fluctuations cannot be derived from equation (3.5). This is due to two factors. First, the definition of the standard deviation (and higher order moments) of the force fluctuations is non-linear. Second, the integrals in equation (3.5) act like a filter in the sense that not all scales participate in creating force fluctuations on a particle. For example, flow scales much smaller than D might cancel out in the integral sense as will be discussed in detail in §3.4. In spite of this observation, a direct relation between flow velocity statistics above the bed or behind an obstacle is often assumed in the literature in order to estimate the intensity of force fluctuations on a particle (cf. Papanicolaou *et al.*, 2002; García, 2008).

In the present simulations it is observed, that the standard deviations for the streamwise and spanwise components of the particle forces are of similar magnitude in both cases F10 and F50 (cf. table 3.2). It is also found that the standard deviation of lift in both cases is roughly half the value of the other two components. Overall the intensity of the fluctuations in case F50 is more than a factor of two larger than in case F10. Thus, the particle force fluctuations in the present case do not seem to scale directly with the intensity of the plane- and time-averaged fluid velocity fluctuations (cf. figure 3.6), since u_{rms} is larger than w_{rms} over most of the flow depth, and especially close to the wall. Furthermore, the difference in the fluid velocity fluctuation intensities between case F10 and F50 is

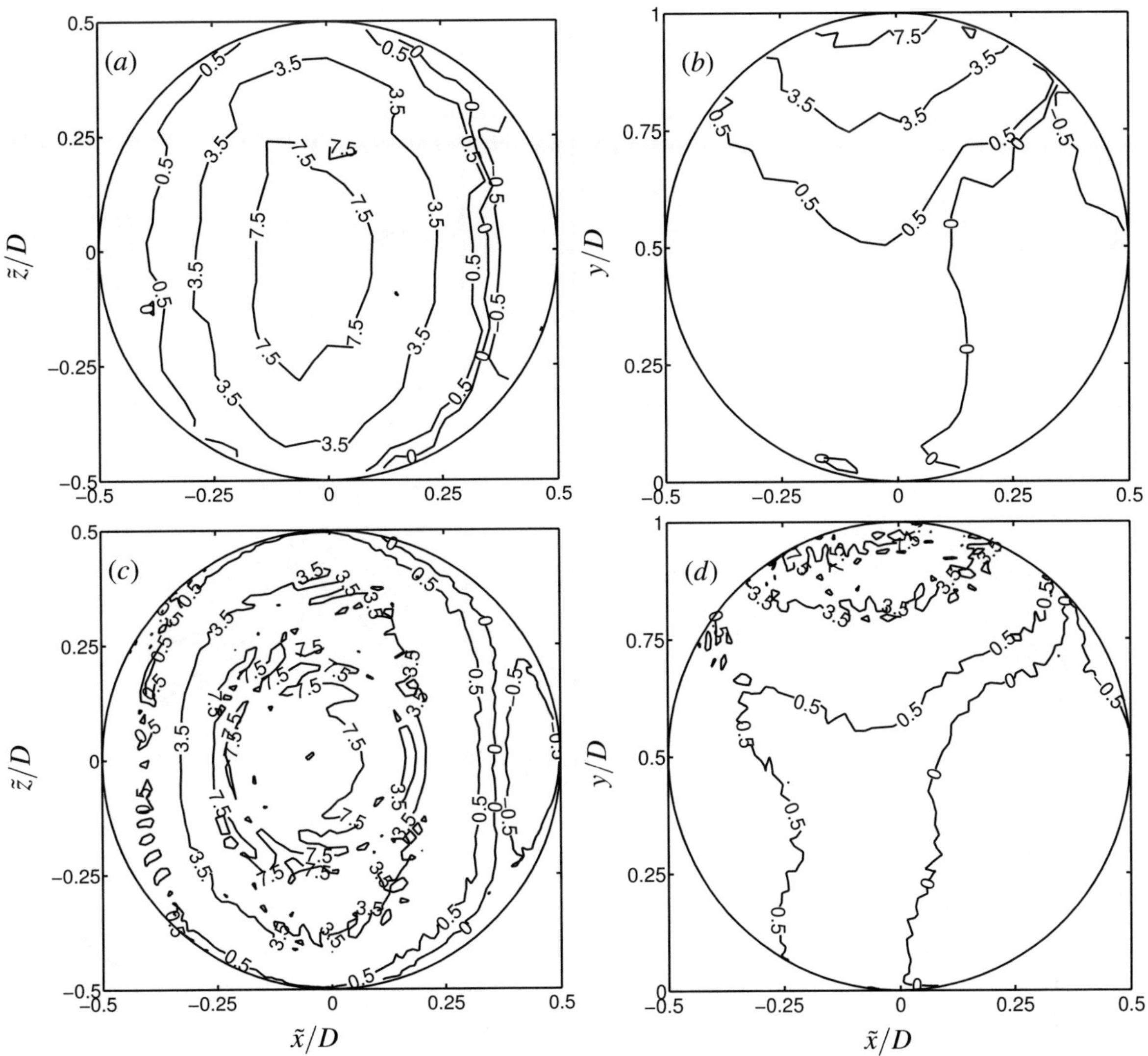

Figure 3.13: Spatial distribution of τ_D, normalised by F_R/A_{sph} in case F10 (a,b) and case F50 (c,d). $\tilde{x}$ and $\tilde{z}$ are the coordinates with respect to the particle centre. The contour lines correspond to [-0.5 0.0 0.5 3.5 7.5].

very small compared to the above stated difference in the particle force fluctuation intensities. It can therefore be concluded that a direct link between fluid and particle force fluctuation intensities cannot be inferred in the present cases.

The results of skewness and kurtosis of the force distributions (cf. table 3.2) are now discussed jointly with the probability density function (pdf) of the particle force fluctuations shown in figure 3.17. For both case F10 and case F50 the highest skewness is obtained for lift, i.e. S_{Fy}. In other words, large positive lift fluctuations are significantly more likely to occur than large negative lift fluctuations. This is clearly visible in figure 3.17(a), where lift events of several standard deviations higher than the mean have a non-negligible probability of occurrence. In case F50, this effect is not as strong as in case F10 (cf. figure 3.17b), and accordingly the value of the skewness S_{Fy} is lower in the former case. The small positive skewness of the drag indicates similarly that instantaneous high drag events are more likely compared to low drag events. For this component, however, the effect appears

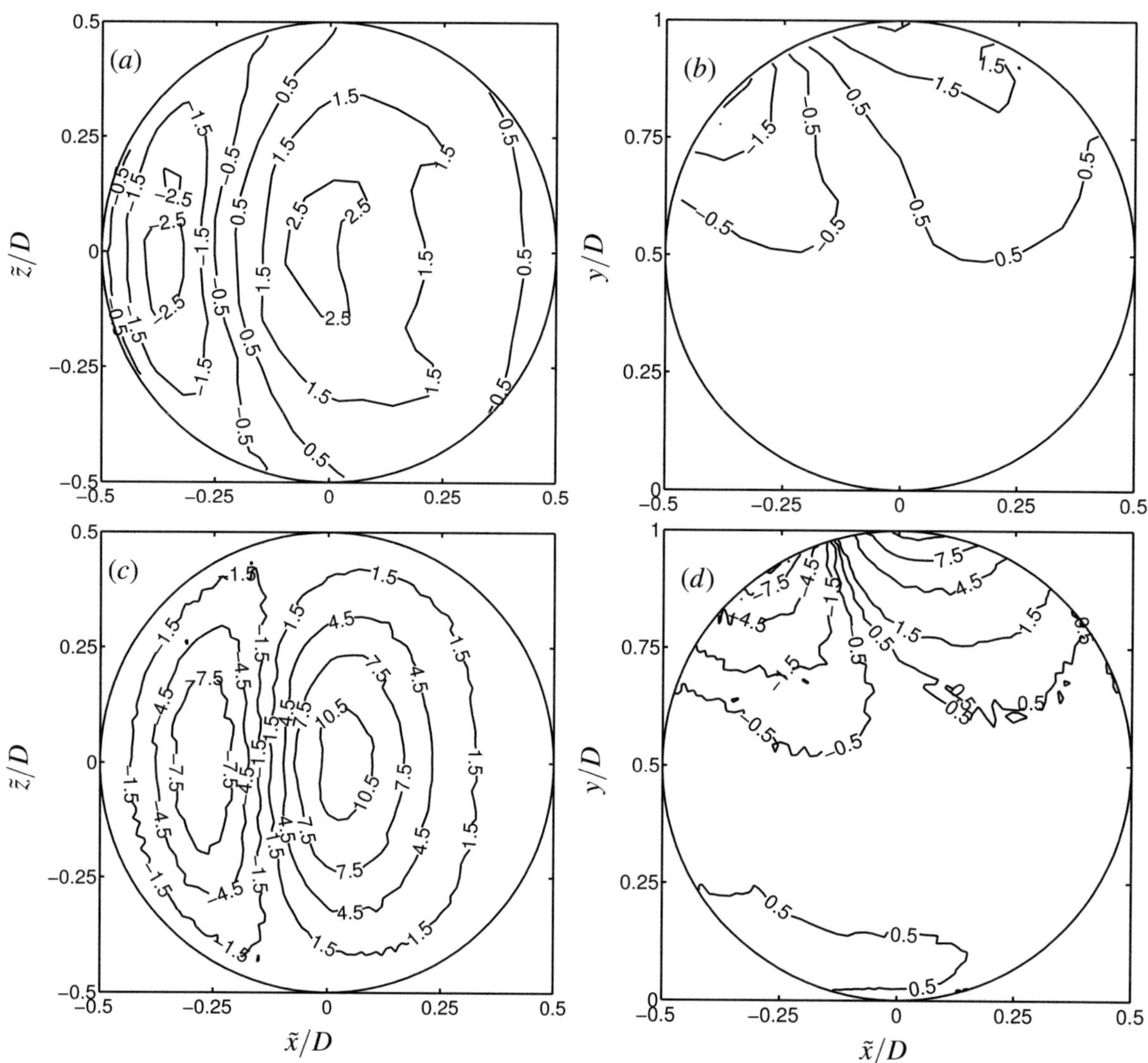

Figure 3.14: As figure 3.13, but for τ_L. The contours lines in (*a*) and (*b*) are at values from -2.5 to 2.5 in steps of 1, in (*c*) and (*d*) at values from -7.5 to 10.5 in steps of 3. In (*d*) additionally the contours at the values -0.5 and 0.5 are shown.

to be much weaker as compared to lift. Finally, symmetry arguments again lead to the conclusion that S_{Fz} should be zero, and this is indeed the case.

The kurtosis of all profiles is rather large, indicating a strong intermittency of the forces, i.e. the pdfs in figure 3.17 exhibit much longer tails than a Gaussian distribution. However, as the spheres become larger the values of skewness and kurtosis approach the Gaussian values of zero and three. This trend might be due to the fact that the force on the particle is an integral quantity, and as mentioned before, small intermittent events might be averaged out. This argument is further elaborated in §3.4.

The results above might be compared to the experiments of Mollinger & Nieuwstadt (1996) on lift fluctuations on a single sphere with $D^+ = 2.9$ positioned on top of a smooth wall. Although their flow configuration is somewhat different (no sheltering effect, turbulent boundary layer) they also report positive values for the skewness ($S_{Fy} = 1.2$) and high values of flatness ($K_{Fy} = 7.0$). Furthermore,

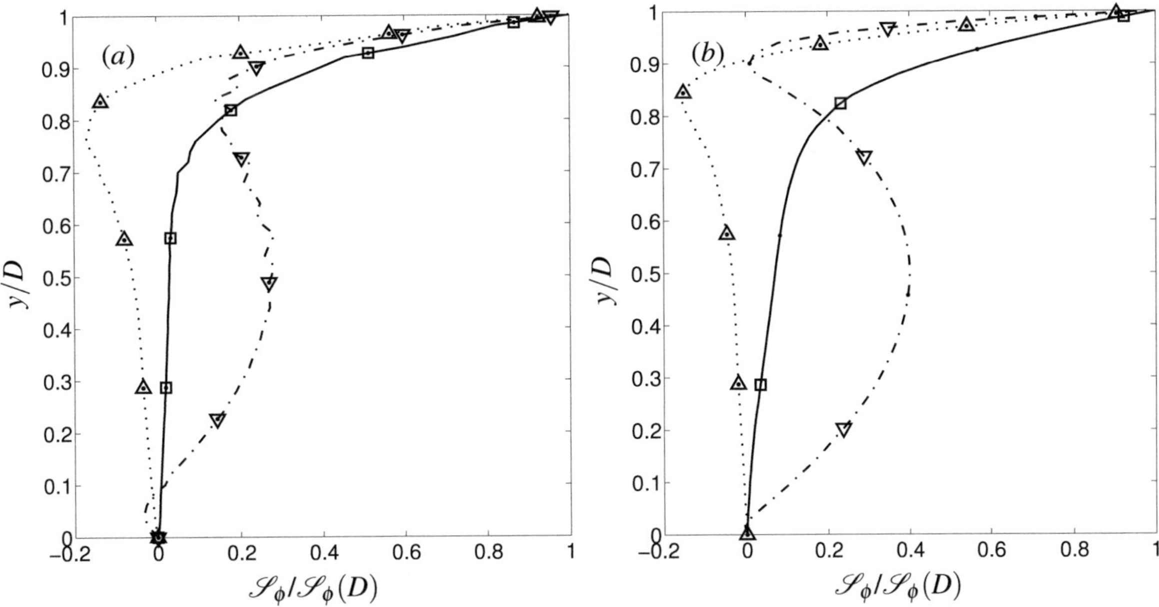

Figure 3.15: Cumulative function $\mathscr{S}_\phi$ of the stress contribution to the mean value of drag, lift and spanwise torque on a particle as in case F10 (*a*) and case F50 (*b*) a function of y/D and normalised by its maximum value, $\mathscr{S}_\phi(D)$. Lines with symbols corresponds to drag($-\square-$), lift ($-\cdot-\bigtriangledown-\cdot-$) and spanwise torque ($\cdots\triangle\cdots$).

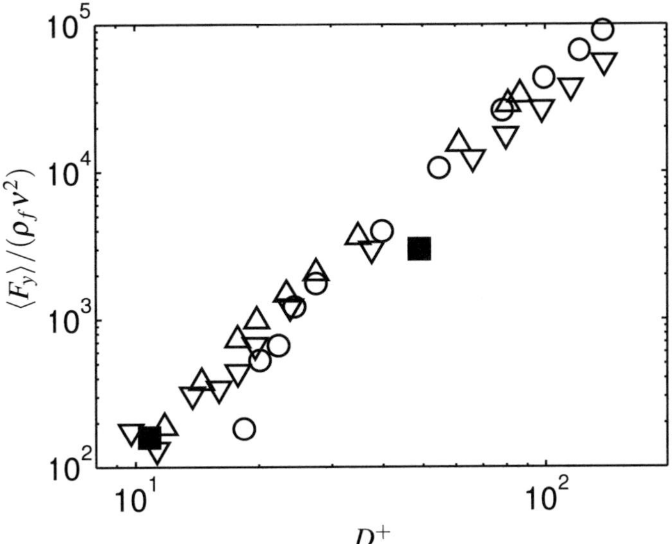

Figure 3.16: Comparison of $\langle F_y\rangle/\left(\rho_f v^2\right)$ as a function of D^+ of case F10 and case F50 (solid symbols) with mean lift on a sphere placed in between roughness elements in a boundary layer by Hall (1988) (open symbols). Symbols correspond to present simulations (■), experiments Hall (1988) with $D_r = 5/3D$ ($\bigtriangledown$), $D_r = D$ ($\bigcirc$), $D_r = 2/3D$ ($\triangle$). Here D_r is the diameter of rods spaced with D_r upstream and downstream of a sphere with diameter D.

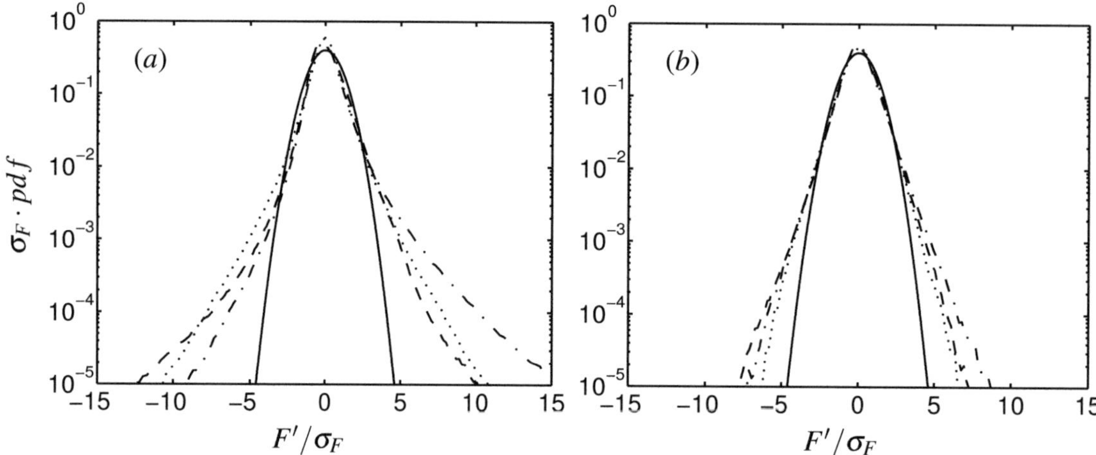

Figure 3.17: Normalised probability density functions of force fluctuations in case F10 (*a*) and case F50 (*b*). Lines show Gaussian distribution (———), F_x'/σ_{Fx} (– – –), F_y'/σ_{Fy} (– · –), F_z'/σ_{Fz} (· · · · · ·).

Case	C_{Tx}	C_{Ty}	C_{Tz}	σ_{Tx}/T_R	σ_{Ty}/T_R	σ_{Tz}/T_R	S_{Tx}	S_{Ty}	S_{Tz}	K_{Tx}	K_{Ty}	K_{Tz}
F10	0.00	0.00	-0.98	0.21	0.04	0.36	0.01	-0.01	-1.04	6.46	6.17	4.72
F50	0.00	0.00	-0.73	0.17	0.11	0.27	-0.01	-0.01	-0.76	3.75	4.91	3.37

Table 3.3: Statistics of torque on particles in case F10 and F50. $C_{Ti} = \langle T_i \rangle / T_R$ is the normalised mean torque component in the x_i-direction, σ_{Ti} is the standard deviation of the torque in x_i direction, S_{Ti} and K_{Ti} are the skewness and kurtosis of the respective torque component.

the pdf of the lift fluctuations in their study is of similar shape to the one obtained in the present case F10. This qualitative agreement suggests that the present results might be relevant to a broader range of flow configurations, e.g. different sphere arrangements or packing densities.

3.4 Statistics of particle torque

The hydrodynamic torque $\mathbf{T}$ on a spherical particle with respect to its centre is defined as

$$\mathbf{T} = \int_\Gamma \mathbf{r}_p \times (\boldsymbol{\tau} \cdot \mathbf{n}) \, \mathrm{d}\Gamma, \tag{3.10}$$

where $\mathbf{r}_p = (r_x^p, r_y^p, r_z^p)$ is the distance vector from the particle centre to an element of the surface Γ. It should be noted that – contrary to the definition of the hydrodynamic force on a particle (3.5) – the pressure does not enter the integral (3.10), since in the present case the differential pressure force $-p^{tot}\mathbf{n}\mathrm{d}s$ on a surface element $\mathrm{d}s$, is always directed towards the particle centre. Based on the reference force F_R given in §3.3 the reference torque is defined as $T_R = F_R r_R$ where r_R is the distance from the particle centre to the virtual wall, $r_R = y_0 - D/2$. The quantity T_R will be used in the following for the normalisation of the various torque related statistical values. A sketch that illustrates the definition of the torque on a particle can be seen in figure 3.12(*a*).

Table 3.3 shows the statistical moments of the torque on the particles. Here, C_{Ti} is the mean torque in the x_i-direction normalised by T_R. Once more, due to symmetry the only non-zero component of the mean torque is expected to be C_T^z. The table shows that negative mean values for the spanwise

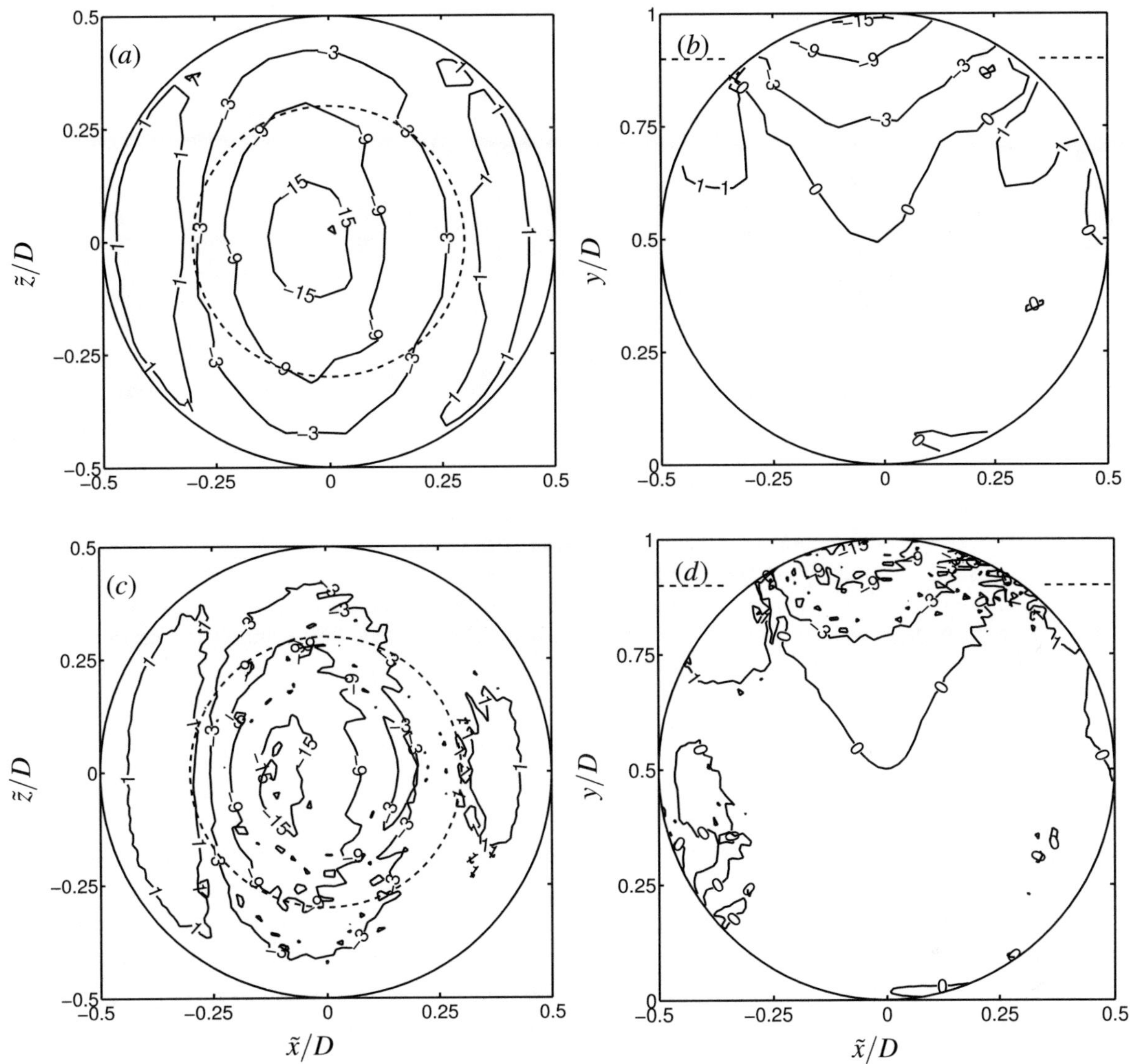

Figure 3.18: Spatial distribution of τ_T, normalised by T_R/A_{sph} in case F10 (*a*,*b*) and case F50 (*c*,*d*). $\tilde{x}$ and $\tilde{z}$ are the coordinates with respect to the particle centre. Contour lines are shown at values of $[-15\ -9\ -3\ 1]$ in all plots. In (*b*) and (*d*) additionally the contour line at zero value is shown. The dashed line indicates the location of $y = 0.9D$.

component are obtained. These negative values of C_T^z are expected for the torque on a particle in positive shear (cf. figure 3.4 and figure 3.9). The torque coefficient as defined above takes values close to -1 for case F10, while it is approximately 25% lower in magnitude in case F50.

To analyse the mean spanwise torque on a sphere in more detail, figure 3.18 shows the distribution of stresses to mean spanwise torque on the sphere's surface,

$$\tau_T = \tau_L r_x^p - \tau_D r_y^p \,. \tag{3.11}$$

In both cases, the distribution of τ_T in figure 3.18 is similar in shape and magnitude. Similar to the distribution of τ_D before (cf. figure 3.13), a shift towards the particle front can be observed for the

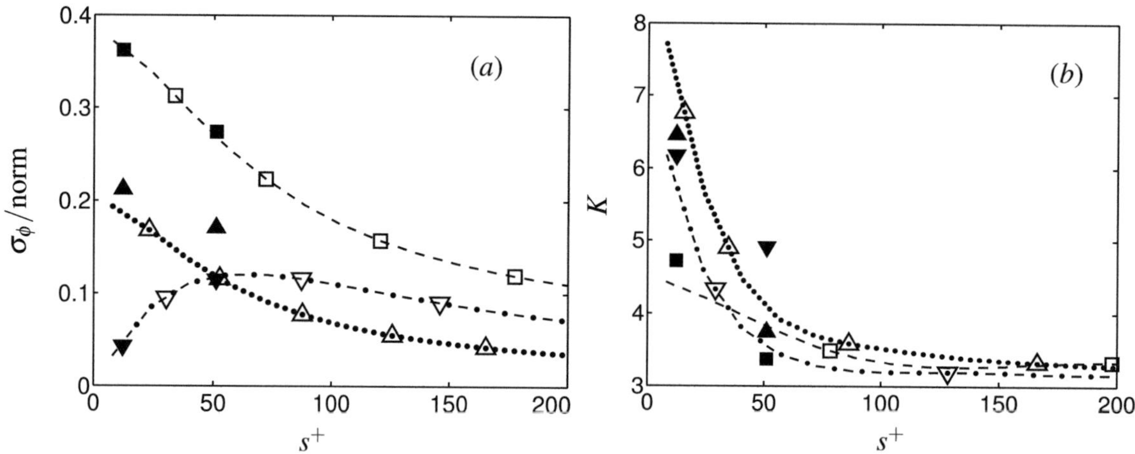

Figure 3.19: Root-mean-square value (*a*) and kurtosis (*b*) of force and torque obtained from the simple model described in the text and of actual values of the corresponding torque obtained on the particles in case F10 and F50. In the former case (smooth wall) the integration is performed over a square wall element of side length s. The open symbols and lines corresponding to $\mathscr{F}_x$ ($--\square--$), $\mathscr{T}_y$ ($-\cdot-\bigtriangledown-\cdot-$), $\mathscr{F}_z$ ($\cdots\triangle\cdots$). The forces are normalised by $\rho_f u_\tau^2 s^2$, torque is normalised by $1/2\rho_f u_\tau^2 s^3$. In the latter case (rough wall) the integration is taken over the particle surface (as given in equation 3.10) with the filled. Full symbols correspond T_x (▲), T_y (▼), T_z (■). Note that in cases F10 and F50 the reference length s is taken as $\sqrt{A_R}$.

minimum values of τ_T near the particle top. This shift is more pronounced in case F50. Negative values of τ_T occur almost exclusively in the upper part of the particle. Thus over most part of the sphere τ_T is positive, but low in magnitude. The cumulative contribution function of τ_T, denoted by $\mathscr{S}_T$ (cf. equation 3.9), is also shown in figure 3.15. The figure reveals that the integration of τ_T in the lower part of the sphere adds up to approximately $-0.15C_{Tz}$ in the vicinity of the virtual wall. In both cases the values of τ_T are predominantly negative for wall-distances above $y \approx 0.8D$, such that $\mathscr{S}_T$ vanishes around $y = 0.9D$. It can therefore be argued, that the net spanwise torque C_{Tz} is generated in the surface area between a wall distance of $y = 0.9D$ and the particle top (i.e. the region highlighted by a dashed line in figure 3.18).

Before turning to the discussion of the torque fluctuations, first a simple model is introduced, which allows us to elucidate some of the characteristics of torque by considering flow structures related to the torque fluctuations on a particle. It should be noted that other authors have previously investigated the relation between flow structures at different scales and the forces/torque exerted upon sediment particles (e.g. Hofland, 2005). Here we employ a somewhat different approach which allows us to use data from a smooth wall flow. In particular, we first analyse drag and torque fluctuations experienced by a square wall-element in channel flow with a geometrically smooth wall, systematically varying the linear dimension of the wall-element. Subsequently, the obtained statistical results are related to the corresponding statistics of the components of the torque on a spherical particle in the fixed sphere simulations.

Analogously to the definition of force and torque on a particle the force in x and z direction on a square surface element in a smooth wall channel with area $A_s = s^2$, $\mathscr{F}_x$ and $\mathscr{F}_z$, can be defined as

$$\mathscr{F}_x = \int_{-s/2}^{+s/2} \int_{-s/2}^{+s/2} \tau_{xy}\big|_{y=0}\, \mathrm{d}x\mathrm{d}z\,, \quad \mathscr{F}_z = \int_{-s/2}^{+s/2} \int_{-s/2}^{+s/2} \tau_{zy}\big|_{y=0}\, \mathrm{d}x\mathrm{d}z\,, \tag{3.12}$$

where s is the side length of the element and $\tau_{ij}\big|_{y=0}$ are the components of the stress tensor at the wall. The torque on the element with respect to its centre, $\mathscr{T}_y$, can be defined as

$$\mathscr{T}_y = \int_{-s/2}^{s/2}\int_{-s/2}^{s/2} \left(r_x^s \, \tau_{zy}\big|_{y=0} - r_z^s \, \tau_{xy}\big|_{y=0} \right) \mathrm{d}x\mathrm{d}z \,, \tag{3.13}$$

where $\mathbf{r}_s$ is the direction vector with respect to the centre of the area element. A sketch that illustrates the definition of the force and torque on a square element in a smooth wall channel can be seen in figure 3.12(*b*).

Drag and spanwise force on the smooth wall element are expected to be mostly affected by velocity scales in streamwise and spanwise direction, respectively, that are of sizes similar or larger than s. The effect of velocity fluctuations at length scales much smaller than s will tend to cancel out, due to the integral quality of the force (cf. equation 3.12). Thus the highest value of force fluctuation should be expected for smallest values of s, as the contribution of the smaller scales is lost for larger values of s. Conversely, due to the cross-product in (3.13) the torque on a smooth wall surface element is mostly affected by wall-normal vortical motions of sizes comparable to s. The effect of much smaller and much larger scales will cancel out or lead to only small values of torque. Thus for small as well as high values of s small values of $\mathscr{T}_y$ are expected. At some intermediate value of s, the characteristics of wall-normal vortical motions should be most efficient in generating torque, leading to maximum values of $\mathscr{T}_y$. Figure 3.19(*a*) supports that hypothesis. It shows the standard deviation of the forces and torque on the surface element, $\sigma_{\mathscr{F}x}$, $\sigma_{\mathscr{F}z}$ and $\sigma_{\mathscr{T}y}$ normalised by $\rho_f u_\tau^2 s^2$ and $1/2\rho_f u_\tau^2 s^3$, respectively. On a square element with $s^+ \approx 70$ torque appears to be most efficiently produced. This value is somewhat larger than the average distance between the low speed and high speed streak close to the wall which is commonly found to be of order $50\nu/u_\tau$. As can be seen in figure 3.19(*b*), the kurtosis of the above quantities monotonically decreases with the size of the surface element indicating that the intermittency of the small scales is larger than that of the large scales.

A direct analogy between this smooth wall model and the force and torque on a particle is not fully justified, as the flow and the geometry are more complex in the present case. However, some of the characteristics of the particle torque statistics obtained for the present cases can be explained with the aid of such a simple model as will be discussed in the following. Please note that only one torque component can be defined for a plane wall element (here $\mathscr{T}_y$), in addition to the two in-plane forces considered ($\mathscr{F}_x$ and $\mathscr{F}_z$). A correspondence with the three torque components on a spherical particle is established when considering the plane wall element as being located at the top (i.e. the pole located at $y = D$) of the particles in case F10 and F50. The component-wise correspondence is then: $\mathscr{F}_x \to -T_z$, $\mathscr{T}_y \to T_y$, $\mathscr{F}_z \to T_x$ (cf. figure 3.12).

The normalised standard deviations of the particle torque components shown in table 3.3 are all non-zero as can be expected. The amplitudes of the fluctuations of the streamwise and spanwise torque components are found to be the largest, while the wall-normal component is significantly weaker. Compared to the small-sphere case (F10), the streamwise and spanwise components are smaller in the large sphere case (F50), by 20% (σ_{Tx}/T_R) and 25% (σ_{Tz}/T_R), respectively. Contrarily, the wall-normal value σ_{Ty}/T_R is significantly larger in case F50 than in case F10 (nearly by a factor of three).

Figure 3.19, which has already been partially discussed above, also shows the second and fourth statistical moments of particle torque fluctuations as a function of particle size. As can be seen, the standard deviation (figure 3.19*a*) of the wall-normal torque on the particles, T_y, matches rather well the standard deviation of the wall-normal torque exerted on a comparable-size square element in the reference smooth wall flow, $\mathscr{T}_y$. In addition, the figure shows that the standard deviation of the spanwise torque, T_z, compares very well to the standard deviation of the drag exerted on a square element in the smooth wall case, $\mathscr{F}_x$. Concerning the streamwise component of particle torque, T_x, it is found that its standard deviation is somewhat larger than the standard deviation of the spanwise force fluctuations in the smooth wall model, $\mathscr{F}_z$; however, both exhibit a similar decreasing trend with increasing values of the length scale. The overall good agreement between fluctuation intensity of forces/torque on an element of a smooth wall and the corresponding torque components of the particle in case F10 and case F50 is interesting for several reasons. First, it suggests that the significant torque fluctuations are generated in a rather limited region around the particle tops where apparently to some extent the analogy with the hydrodynamic action on a wall-parallel square element holds. In particular, the present simulations F10 and F50 provide two data points in the hydraulically smooth and transitionally rough flow regime, which are fully consistent with the existence of a length scale/particle size of maximum wall-normal torque generation, as suggested by the simplified model. Secondly, if the above analogy is accepted, then it implies that the response of the particle torque fluctuations to the near-wall turbulent flow can indeed be described as a selective filtering effect, mainly characterised by a single length scale (the particle diameter).

Normalised pdfs of the particle torque fluctuations are shown in figure 3.20. It can be seen that the curves for all three torque components in both cases F10 and F50 approximately match the curves of the corresponding force/torque components of the smooth wall model (evaluated with a side-length s matching the respective length $\sqrt{A_R}$), thereby further corroborating the analogy. Concerning the shape of the particle torque pdfs themselves, it is observed that the two symmetric components (streamwise T_x and wall-normal T_y) have significantly longer tails than a Gaussian function, and consequently exhibit higher than Gaussian values of kurtosis (cf. table 3.3). The kurtosis is found to decrease with increasing particle size, consistent with the above filtering argument (also cf. figure 3.15*b*).

The pdf of the fluctuations of the spanwise component of particle torque, T_z, is clearly asymmetric with a pronounced negative skewness. Now, it is well established that the pdf of streamwise velocity fluctuations u' in smooth wall channel flow is positively skewed close to the wall (Kim *et al.*, 1987; Jiménez & Hoyas, 2008). In the limit of a wall-element with vanishing size, the pdf of $\mathscr{F}_x$ is directly related to the pdf of the streamwise velocity fluctuations just above the wall. Since, as found above, the particle torque component T_z behaves similarly as the smooth wall force $-\mathscr{F}_x$ (note the changed sign), the observed negative skewness of the former is consistent with the positively skewed streamwise velocity pdf previously found in smooth wall channel flow.

It should be noted that the analogy drawn between shear forces on a square element of a smooth wall channel flow and the corresponding hydrodynamic torque components of spherical particles in the roughness layer can be expected to lose its appeal in the fully rough regime. In that case, which is outside the scope of the present study, pressure induced forces might by far outweigh viscous forces. Therefore, the main utility of the proposed simple model is presumably limited to the regime of transitional roughness.

3.5 Time scales of force and torque fluctuations

3.5.1 Introduction

In the present context of flow over a rough wall several time scales based on length and velocity scales can be defined. The length scales considered in the following are the particle diameter, D, the effective flow depth, $h = H - y_0$, and the viscous length scale, $\delta_\nu = \nu/u_\tau$. The velocity scales considered are the friction velocity, u_τ, and the bulk velocity based on h, U_{bh}. From the three length scales and two velocity scales, six time scales are defined and considered in the following, namely D/u_τ, D/U_{bh}, h/u_τ, h/U_{bh}, ν/u_τ^2 and $\nu/(u_\tau U_{bh})$. The ratios of the considered length, velocity and time scales

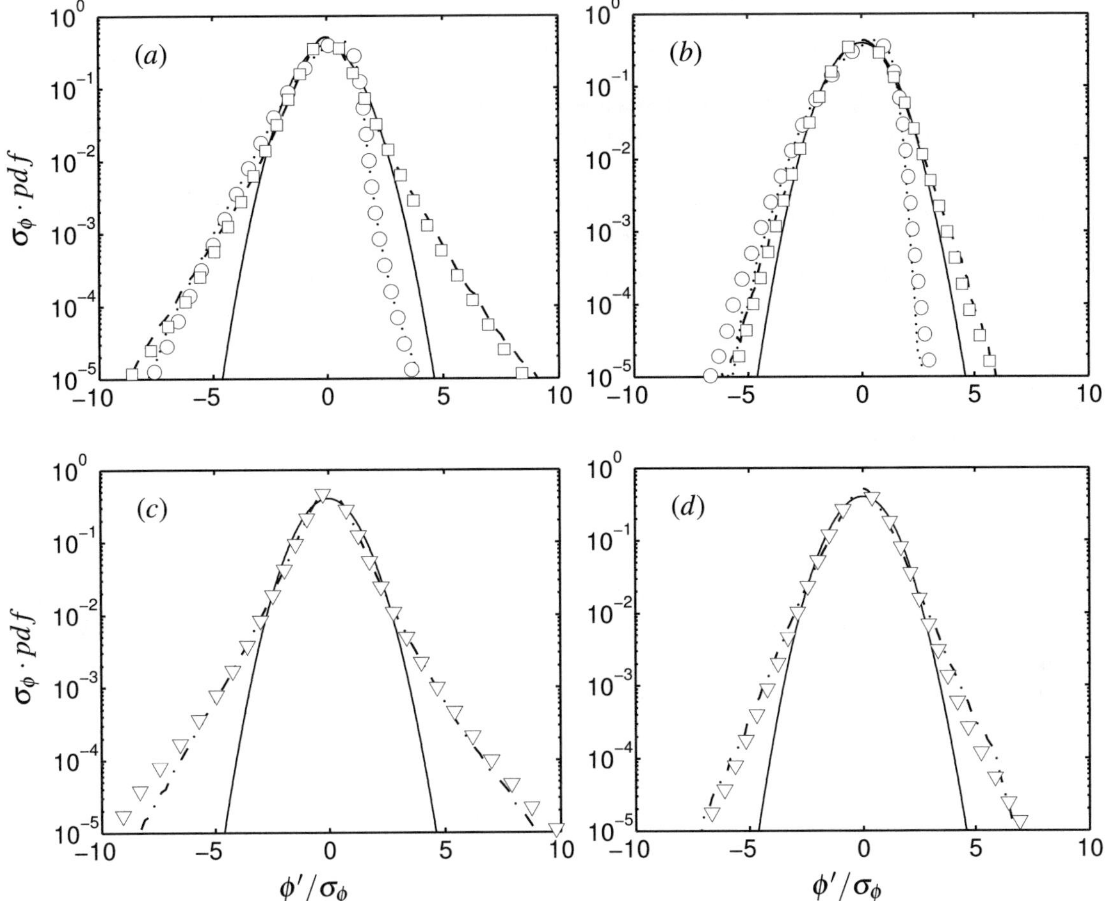

Figure 3.20: Normalised pdfs of the quantities for which statistical moments have been shown in figure 3.19. *(a,c)* Case F10 compared to simple model with $s^+ = 12$. *(b,d)* Case F50 compared to simple model with $s^+ = 52$. *(a,b)* Lines and symbols correspond to T_x'/σ_{Tx} (– – –), T_z'/σ_{Tz} (······), $-\mathcal{F}_x'/\sigma_{\mathcal{F}x}$ (○) (please note the negative sign), $\mathcal{F}_z'/\sigma_{\mathcal{F}z}$ (□). *(c,d)* Lines and symbols correspond to T_y'/σ_{Ty} (– · –), $\mathcal{T}_y'/\sigma_{\mathcal{T}y}$ (▽). The solid line (———) corresponds to a Gaussian distribution.

	D	h	δ_v	U_{bh}	u_τ
F50 / F10	3.29	0.90	0.73	1.12	1.37

Table 3.4: Ratios of length scales and velocity scales in case F50 to those in case F10.

between case F10 and case F50 are given in table 3.4 and 3.5. They show, that the ratio between the scales is of order one in all cases. The largest ratio can be found for length and time scales utilising D.

In the following the time scales of force and torque fluctuations are studied by correlation functions in time defined as

$$R_{\phi\psi}(\tau) = \frac{1}{t_1 - t_0} \int_{t0}^{t1} \phi(t+\tau)\, \psi(t)\, \mathrm{d}t\,, \tag{3.14}$$

where $\phi(t)$ and $\psi(t)$ are two scalar quantities as a function of time, t, such as components of force or torque on a particle or on a smooth wall surface element, τ is the time lag and t_0 and t_1 define a time

	D/u_τ	D/U_{bh}	h/u_τ	h/U_{bh}	ν/u_τ^2	$\nu/(u_\tau U_{bh})$
F50 / F10	2.39	2.94	0.65	0.80	0.73	0.65

Table 3.5: Ratios of time scales in case F50 to time scales in case F10.

	$\tau_m U_{bh}/h$			$\tau_\ell U_{bh}/h$			$\tau_{\min} U_{bh}/h$		
	$\mathscr{F}_x'$	$\mathscr{T}_y'$	$\mathscr{F}_z'$	$\mathscr{F}_x'$	$\mathscr{T}_y'$	$\mathscr{F}_z'$	$\mathscr{F}_x'$	$\mathscr{T}_y'$	$\mathscr{F}_z'$
$s^+ = 11$	0.39	0.21	0.28	1.78	0.56	1.22	7.05	2.25	6.75
$s^+ = 46$	0.48	0.28	0.40	2.10	0.76	1.71	7.10	2.95	7.05
$s^+ = 138$	0.84	0.58	0.77	2.38	1.39	2.48	7.40	5.35	7.50

Table 3.6: Time scales of force and torque fluctuation on square smooth wall surface element with side length, $s^+ = 11$, 46 and 138 normalised by outer flow scales. Considered time scales are the micro time scale, τ_m, the integral time scale, τ_ℓ, and the time-lag related to the first local minimum, $\tau_{\min}$.

interval. To increase the quality of the correlation defined by (3.14) sample averaging is carried out over the number of considered particles, N_p, or over the number of considered smooth wall surface elements.

Three time scales related to auto-correlation functions, $R_{\phi\phi}(\tau)$, will be used below. The time lag $\tau_{\min}$, defined as the time lag related to the first local minimum. The integral time scale, τ_ℓ, as a measure of the largest time scales. This time scale is defined in analogy to the integral length scale (see O'Neill *et al.*, 2004, Tritton, 1988 for a discussion) as

$$\tau_\ell = \frac{1}{\sigma_\phi^2} \int_0^{\tau_{\min}} R_{\phi\phi}(\tau) \mathrm{d}\tau, \tag{3.15}$$

where σ_ϕ is the standard deviation of the quantity ϕ.

The micro-scale, τ_m, is introduced as a measure of the smallest times scales. Motivated by the Taylor[1] micro-scale (e.g. Pope, 2000, p. 199) it is defined as

$$\tau_m = \frac{\sqrt{2}\sigma_\phi}{\sqrt{-\ddot{R}_{\phi\phi}(0)}}, \tag{3.16}$$

where $\ddot{R}_{\phi\phi}(\tau)$ is the second derivative of $R_{\phi\phi}(\tau)$ with respect to τ.

3.5.2 Time scales of force and torque fluctuations on a smooth wall surface element

Here, the auto-correlation functions of force and torque on a square smooth wall surface element are considered based on the data of reference case S180 (cf. appendix B). The streamwise force, $\mathscr{F}_x$, the spanwise force, $\mathscr{F}_z$ and the wall-normal torque, $\mathscr{T}_y$ are defined by (3.12) and (3.13). Three different side length of the square surface element are used, $s^+ = 12$, $s^+ = 52$ and $s^+ = 209$. The auto-correlation, $R_{\phi\phi}(\tau)$, of the signals are defined by (3.14) employing $\phi = \psi$. Figures 3.21 and 3.22 show auto-correlations of the streamwise force fluctuation, $\phi = \mathscr{F}_x'$, the spanwise force fluctuation, $\phi = \mathscr{F}_z'$, and the wall-normal torque fluctuation, $\phi = \mathscr{T}_y'$, of the in panel (*a*), (*b*) and (*c*) respectively. Note, that for $s^+ = 209$ the side length of the square element is larger than h, i.e. $h^+ = \mathrm{Re}_\tau = 183 < s^+ = 209$.

[1] Sir Geoffrey Ingram Taylor OM, English physicist and mathematician, ⋆7 March 1886 †27 June 1975

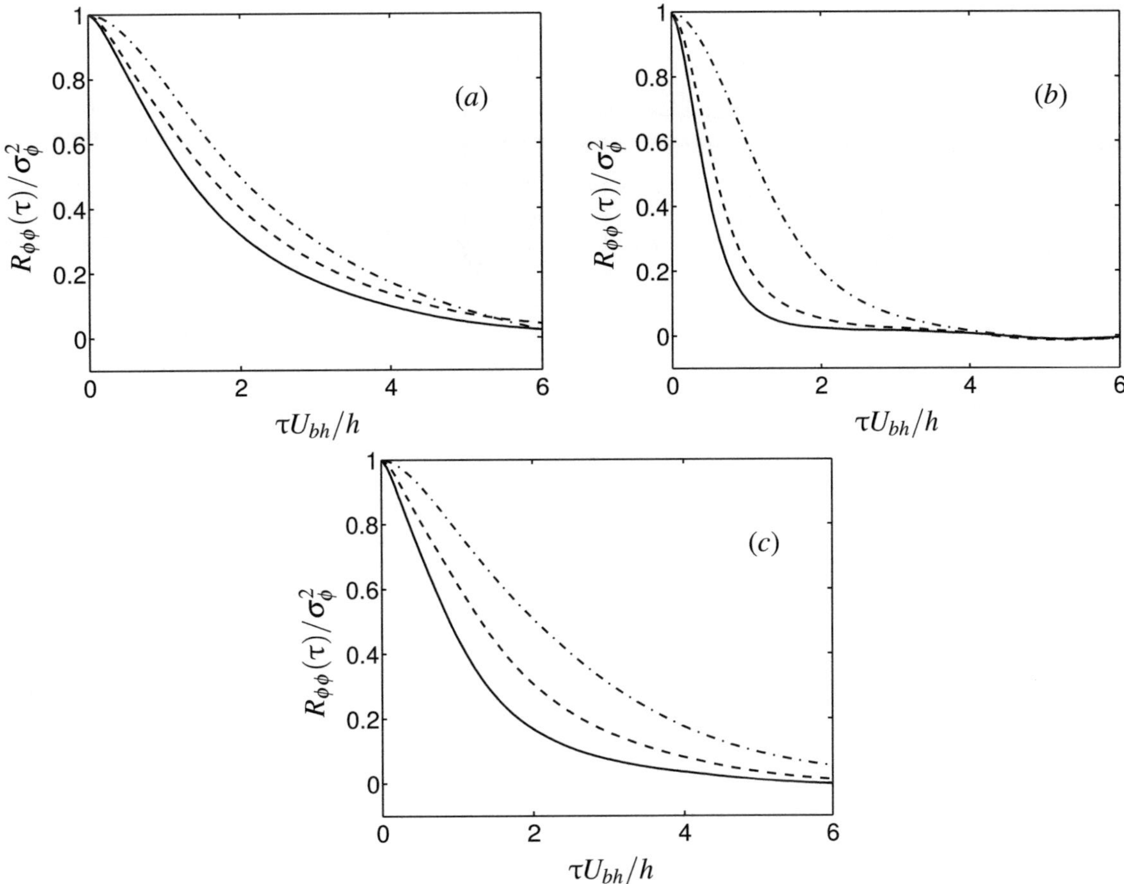

Figure 3.21: Auto-correlation, $R_{\phi\phi}(\tau)/\sigma_\phi^2$, as a function of time lag $\tau U_{bh}/h$ with ϕ as the force or torque fluctuation on a square smooth wall surface element. Panels show (a) streamwise force fluctuation $\mathscr{F}_x'$, (b) spanwise force fluctuation $\mathscr{F}_z'$, (c) torque fluctuation $\mathscr{T}_y'$. Lines show results of square surface elements with side lengths $s^+ = 12$ (———), $s^+ = 52$ (– – –), $s^+ = 209$ (– · –).

In the following the auto-correlations are discussed jointly with respective values of τ_m, τ_ℓ and $\tau_{\min}$ provided in table 3.6.

The profiles of the auto-correlation of force and torque fluctuation on the smooth wall element are positive and decrease monotonically for $0 \leq \tau U_{bh}/h < 5$ (figure 3.21). As might be guessed from figure 3.21, the time micro-scale, $\tau_m U_{bh}/h$, as well as the integral time scale, $\tau_\ell U_{bh}/h$, increase with increasing s (table 3.6). The increase in $\tau_m U_{bh}/h$ might be put into context with the filtering argument utilised in §3.4 to link flow structures to force and torque on a smooth wall surface element. There it was argued, that the length of the smallest flow scales which affect force and torque fluctuations, increases with s^+ for sufficiently large s^+. Following this idea the increase of $\tau_m U_{bh}/h$ of force and torque on a square smooth wall surface element with the side length could be interpret as that small flow motions with small time scales cease influence on the force and torque fluctuation with increasing side length of the element. This would then imply a relation between the length scale of a flow structure to its time scale, which is indeed often assumed to be the case.

More difficult to interpret is the increase of $\mathscr{F}_x'$ and $\mathscr{F}_z'$ with s^+ (table 3.6). Earlier it was argued, that the effect of the large scales on $\mathscr{F}_x'$ and $\mathscr{F}_z'$ is independent of s and thus it might be surprising that

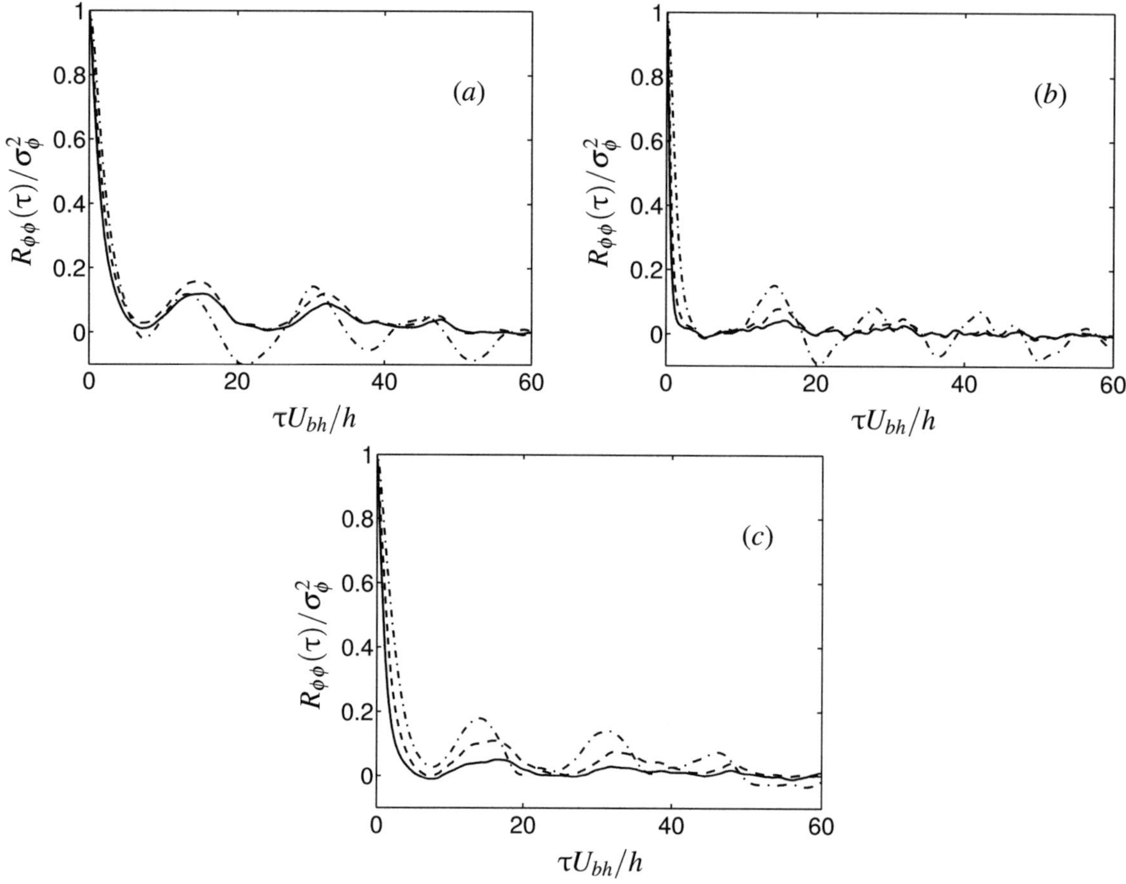

Figure 3.22: As in figure 3.21 but for $0 \leq \tau U_{bh}/h \leq 60$.

$\tau_\ell U_{bh}/h$ increases with s^+. However, recall that the standard deviation of drag and spanwise force on the square element decreases with increasing s (cf. figure 3.19*a*). Thus, the relative contribution to the auto-correlation normalised by the standard deviation increases (cf. figure 3.21) even when the absolute contribution of the large scales to the auto-correlation is constant.

The time scales in table 3.6 are of order one when scaled in outer units, the measure of the small as well as the measure of the large time scales. This reflects a lack of scale separation between small and large scales in the present case.

Figure 3.22 shows fluctuations with a period of approximately $\mathscr{T} U_{bh}/h \approx 15$ which lead to non-negligible correlation coefficients for large time separations. This period can be linked to the effect of the streamwise periodicity on the largest flow structures. An estimation of the period of the largest length scales of motion for the given streamwise periodicity can be obtained by assuming a convection velocity of $10u_\tau$ for the largest scales at the wall (cf. del Álamo & Jiménez, 2009). This leads to a value of $L_x U_{bh}/(10u_\tau h) \approx 19$, which indicates that the large scale fluctuation is a result of the periodicity in streamwise direction and might bias the auto-correlation of force and torque on a smooth wall surface element.

Similar conclusions as from the auto-correlations, $R_{\phi\phi}(\tau)$, above can be drawn from the spectra in time, $\widehat{R}_{\phi\phi}(\omega)$, where ω is the frequency of the time signal $\phi(t)$. In case of infinite or periodic time signals $\widehat{R}_{\phi\phi}(\omega)$ can be defined as the Fourier transform of the auto-correlation or alternatively

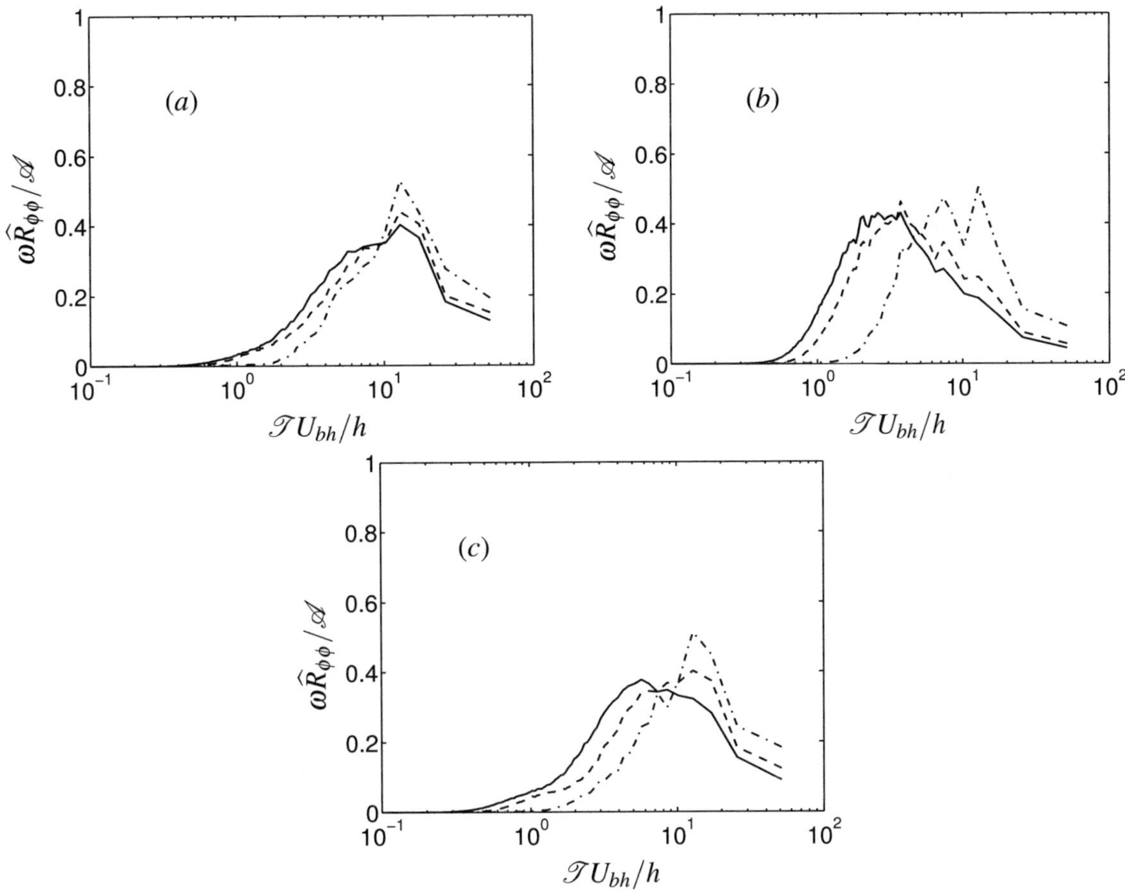

Figure 3.23: Pre-multiplied spectra, $\omega\widehat{R}_{\phi\phi}$, of force and torque fluctuations on square smooth wall element as a function of period $\mathscr{T}U_{bh}/h$. The spectra are normalised by $\mathscr{A}$ as defined in the text. Panels and line styles as in figure 3.21.

as $\widehat{R}_{\phi\phi}(\omega) = \widehat{\phi}(\omega)\widehat{\phi}^*(\omega)$, where $\widehat{\phi}(\omega)$ is the Fourier transform of $\phi(t)$, and $\widehat{\phi}^*(\omega)$ the conjugate complex of $\widehat{\phi}$ (see for example appendix D in Pope, 2000, for an introduction). Thus, the spectra contains the same information as the auto-correlation but presents it in a different form. Here, the time signals considered are finite and non-periodic and $\widehat{R}_{\phi\phi}(\omega)$ is approximated by the method of Welch (1967) with 50% overlap and applying a Hamming[1] window (cf. Oppenheim & Schafer, 1989, pp. 447-448) to the original signal ϕ, prior of transferring it into spectral space.

Figure 3.23 shows the pre-multiplied spectra $\omega\widehat{R}_{\phi\phi}$ of $\mathscr{F}_x'$, $\mathscr{F}_z'$ and $\mathscr{T}_y'$ in panel (a), (b) and (c), respectively. The spectra are shown as a function of period $\mathscr{T}U_{bh}/h$ and are normalised by $\mathscr{A} = \int_{-\infty}^{-\infty} \widehat{R}_{\phi\phi}\mathrm{d}\omega$, such that the integral of the spectra over the frequency returns a value of unity. This corresponds to the normalisation of the auto-correlation function by σ_ϕ^2, i.e. $R_{\phi\phi}(0)/\sigma_\phi^2 = 1$ and emphases their frequency content.

In all three panels the value of $\omega\widehat{R}_{\phi\phi}/\mathscr{A}$ for small periods $\mathscr{T}U_{bh}/h$ decreases with increasing s^+ which is in agreement with the findings discussed above. Similarly, the value of $\omega\widehat{R}_{\phi\phi}/\mathscr{A}$ for large periods $\mathscr{T}U_{bh}/h$ increases with s^+. The figures reveal that the most contribution to $\mathscr{F}_{x\mathrm{rms}}$ and $\mathscr{T}_{y\mathrm{rms}}$ is

[1]Richard Wesley Hamming, American mathematician, ⋆ 11 February 1915 † 7 January 1998

	$\tau_m u_\tau^2/\nu$			$\tau_\ell u_\tau^2/\nu$			$\tau_{\min} u_\tau^2/\nu$		
Case	T_x'	T_y'	T_z'	T_x'	T_y'	T_z'	T_x'	T_y'	T_z'
F10	5.9	7.8	10	7.0	11	22	57	62	88
F50	8.5	10	11	9.4	13	20	63	77	154

Table 3.7: Time scales of torque fluctuations on a particle in case F10 and F50 normalised by ν/u_τ^2. Considered time scales are the micro time scale, τ_m, the integral time scale, τ_ℓ, and the time-lag related to the first local minimum $\tau_{\min}$.

$\mathscr{T}_{\max} u_\tau^2/\nu$	T_x'	T_y'	T_z'	$\mathscr{T}_{\max} U_{bh}/h$	T_x'	T_y'	T_z'
F10	37	54	175	F10	3.0	4.3	14
F50	50	57	99	F50	2.6	3.0	5.3

Table 3.8: Period of local maximum in pre-multiplied spectra of torque fluctuations, $\mathscr{T}_{\max}$, normalised by outer and inner flow scales.

contained in time scales with periods in the range of $2 < \mathscr{T} U_{bh}/h < 20$. It is interesting to note, that for $\mathscr{F}_z'$ with $s^+ = 12$ and $s^+ = 52$, the maximal contribution stems from somewhat smaller time scales. The maxima of the pre-multiplied spectra of $\mathscr{F}_x$ in figure 3.23(*a*) is located around $\mathscr{T} U_{bh}/h = 19$. This period compares to those discussed in the auto-correlation of figure 3.22 and has been linked to the streamwise periodicity of the current setup. The maxima in figure 3.23(*b*,*c*) for $s^+ = 12$ is located at lower values of $\mathscr{T} U_{bh}/h$, but shift to comparable values as s increases.

To conclude, in the present simulations the time scales of force and torque fluctuations on a square smooth wall surface element are of the order of the outer flow time scales. As expected the range of existing time scales is small and lack a distinct separation of scales. Auto-correlation and spectra reveal that small time scales lose relevance in favour of large time scales as the side length of the element increases. This supports the idea raised in the context of the filtering argument discussed in 3.4, that flow motions related to small length and time scales cease influence on force and torque fluctuations on a smooth wall surface element when the side lengths is increased. For high values of s^+, the limitation of the largest scales by the streamwise periodicity becomes more critical and might lead to a bias in the present smooth wall cases.

3.5.3 Time scales of torque fluctuations on a particle

In the following, the auto-correlation, $R_{\phi\phi}(\tau)/\sigma_\phi^2$, of particle torque fluctuation in case F10 and case F50 as defined by (3.14) are discussed. The auto-correlations are computed from the particle data collected during run-time of the simulations. As in §3.4 the results are compared to those of the auto-correlation of force and torque on a square smooth wall surface element of side length $s^+ = 12$, i.e. the components of torque on a particle are compared to smooth wall results as follows: $\mathscr{F}_x \to -T_z$, $\mathscr{T}_y \to T_y$, $\mathscr{F}_z \to T_x$. Figure 3.24(*a*,*b*,*c*) shows the auto-correlation of T_x', T_y' and T_z' jointly with the auto-correlation of $\mathscr{F}_z'$, $\mathscr{T}_y'$ and $\mathscr{F}_x'$ for $s^+ = 12$ respectively. The time lag, τ, is normalised by ν/u_τ^2 for which good agreement between the curves is obtained. In particular the curves of T_x' and T_z' of case F10 in panel (*a*) and (*c*) overlap with the smooth wall results $\mathscr{F}_z'$, and $\mathscr{F}_x'$, respectively. Although the differences are somewhat larger in figure 3.24(*b*), they can be considered small. The correlations of case F50 exceed the values of case F10 in figure 3.24(*a*,*b*) which agrees to the behaviour of the respective components in the smooth wall case when increasing s^+ from 12 to 52 (cf. figure 3.23*a*,*c*).

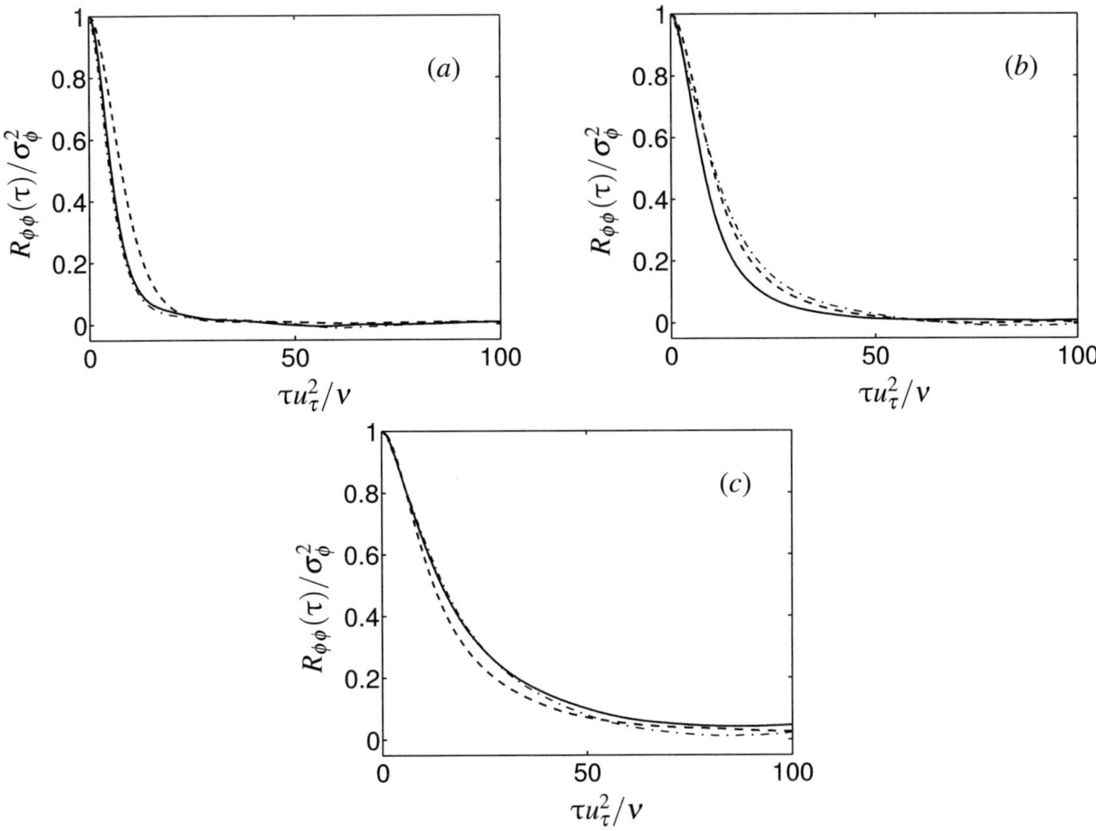

Figure 3.24: Auto-correlation of, $R_{\phi\phi}(\tau)/\sigma_\phi^2$, of particle torque fluctuation as a function of time lag, $\tau u_\tau^2/\nu$. The panels show the results of (*a*) streamwise torque fluctuation, T_x', (*b*) wall-normal torque fluctuation, T_y', and (*c*) spanwise torque fluctuation, T_z'. Lines show results of case F10 (———), case F50 (– – –). Additionally, the results of the smooth wall reference case with $s^+ = 12$ for $\mathscr{F}_z'$, $\mathscr{T}_y'$ and $\mathscr{F}_x'$ are provided as a reference in panel (*a*), (*b*) and (*c*), respectively (– · –).

Here however, in the case of auto-correlation of torque on a particle the effect seems to be less strong. The auto-correlation functions of F_z in figure 3.24(*c*) do not follow this trend, i.e. the correlation is smaller in case F50 than it is in case F10. This is even more pronounced when scaled in outer units (plot omitted). Table 3.7 provides the time scales, τ_m, τ_ℓ and $\tau_{\min}$ of the profiles in figure 3.24. Similar to the force fluctuations, τ_m increases for all torque fluctuations from case F10 to F50 when scaled in viscous units. As can be seen the values of $\tau_\ell u_\tau^2/\nu$ of T_x' and T_y' increase from case F10 to F50, while the value for T_z' decreases.

Some more insight can be gained from the pre-multiplied spectra of the torque fluctuations shown in figure 3.25. The spectra are computed and normalised similar to figure 3.23, but here ν/u_τ^2 is used as reference time scale. In agreement with the filter argument the value $\omega\widehat{R}_{\phi\phi}/\mathscr{A}$ of T_x' and T_y' decreases from case F10 to case F50 for small $\tau u_\tau^2/\nu$ and the contribution of time scales to the root-mean-square of particle torque fluctuations shifts to larger values. For large $\tau u_\tau^2/\nu$ the spectrum of case F50 is found to be higher or equal than the spectrum of case F10, which agrees to the trend observed in the smooth wall case (cf. figure 3.23). For the T_z' the spectra appear to be of similar shape. A distinct difference is that the spectra of case F10 is less smooth and exhibits a pronounced maxima at a period, $\mathscr{T} u_\tau^2/\nu = 175$. This value is close to twice as large as that in case F50 ($\mathscr{T} u_\tau^2/\nu = 99$),

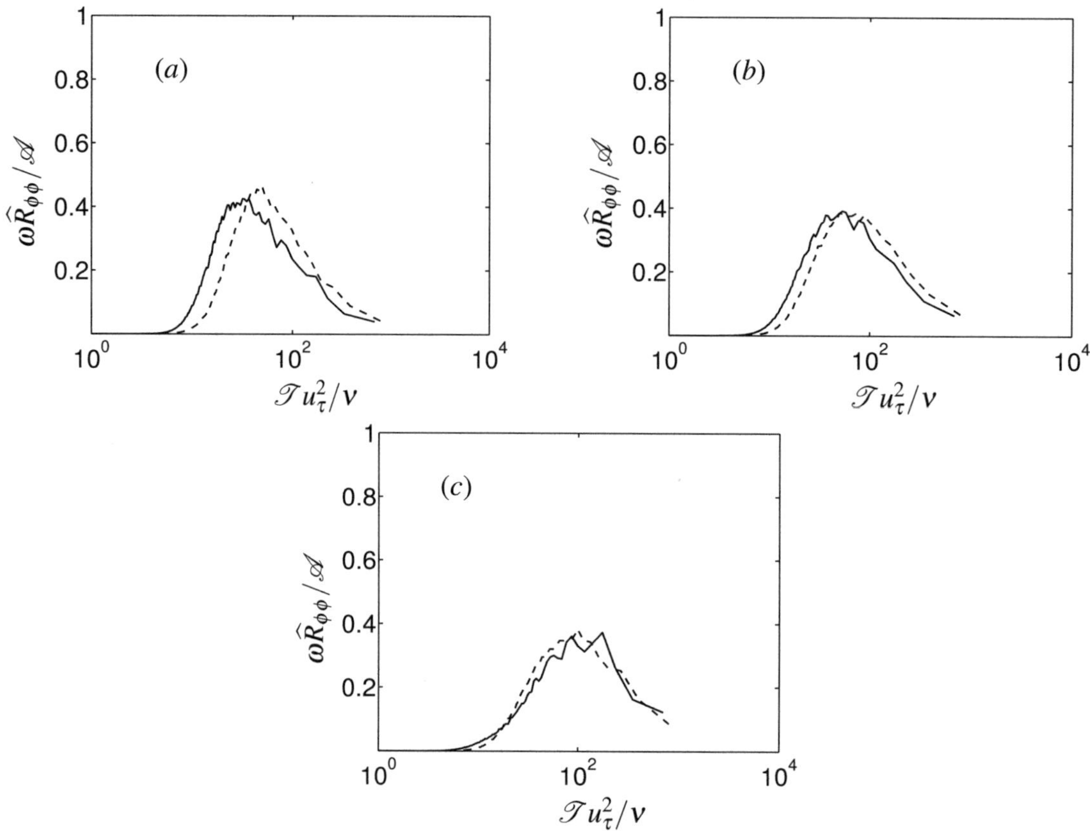

Figure 3.25: Pre-multiplied spectra $\omega\widehat{R}_{\phi\phi}$ of particle torque fluctuations as a function of period $\mathscr{T}u_\tau^2/\nu$. Spectra are normalised by $\mathscr{A}$ as defined in the text. Panels and line style as in figure 3.24.

and might be a bias by the restriction of the domain size in streamwise direction. Also, when scaled in outer units $\mathscr{T}_{\max}U_{bh}/h = 14$, and thus is comparable to the period related to the effect of the domain size limitation in the smooth wall case. The periods related to the maxima in figure 3.25, $\mathscr{T}_{\max}$, are provided in table 3.8, in outer and inner units. It should be pointed out, that these maxima represent a broad range of periods. The values show the trend which was already discussed, i.e. while a shift to somewhat higher values of $\mathscr{T}_{\max}u_\tau^2/\nu$ can be observed for T_x' and T_y' from case F10 to F50 (cf. figure 3.25*a*,*b*), the value of $\mathscr{T}_{\max}u_\tau^2/\nu$ decreases from case F10 to F50 for T_z'. Additionally the table shows that when scaled in outer flow units $\mathscr{T}_{\max}$ decreases for all components of torque fluctuations.

3.5.4 Time scales of force fluctuations on a particle

The auto-correlation, $R_{\phi\phi}(\tau)/\sigma_\phi^2$, of particle force fluctuation in case F10 and case F50 as defined by (3.14) are provided in figures 3.26 and 3.27. As in the case of torque on particles, the auto-correlations are computed from the particle data collected during run-time of the simulations. For comparison, the auto-correlation of drag fluctuation, $\mathscr{F}_x'$, and spanwise force fluctuation, $\mathscr{F}_z'$, on a smooth wall surface element with $s^+ = 12$ are included in panels (*a*) and (*c*). In contrast to the auto-correlation of torque on particles, it is found that the characteristics of the auto-correlations of force differ between smooth wall case, case F10 and case F50 (figure 3.26). The profiles of $\mathscr{F}_x'$ and $\mathscr{F}_z'$ are positive and decrease monotonically in the given range of figure 3.26, while in case F50 the profiles of F_x', F_y' and F_z' exhibit

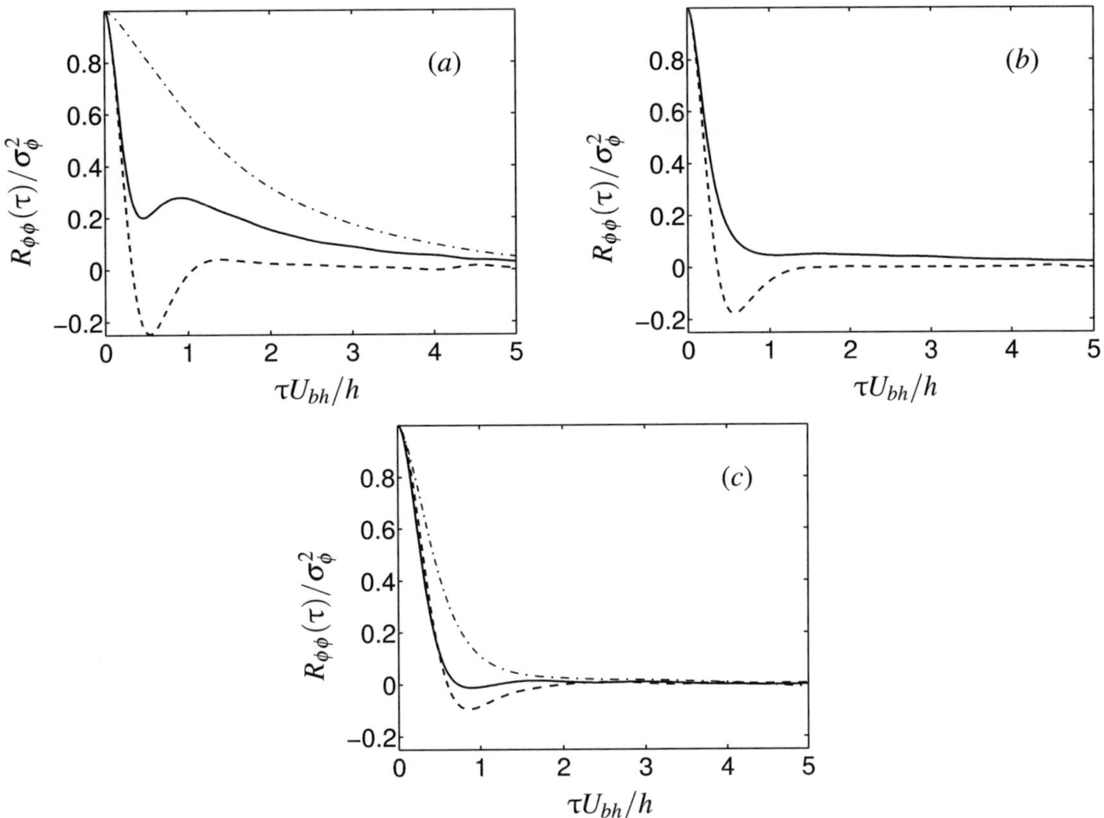

Figure 3.26: Auto-correlation, $R_{\phi\phi}(\tau)/\sigma_\phi^2$, of particle force fluctuation as a function of time lag, $\tau U_{bh}/h$. The panels show results of (*a*) drag fluctuation, F_x', (*b*) lift fluctuation, F_y', and (*c*) spanwise force fluctuations, F_z'. Lines show results of case F10 (———), case F50 (– – –). Additionally, the results of the smooth wall reference case with $s^+ = 12$ for $\mathscr{F}_x'$ and $\mathscr{F}_z'$ are provided as a reference in panel (*a*) and (*c*), respectively (– · –).

	$\tau_m U_{bh}/h$			$\tau_\ell U_{bh}/h$			$\tau_{\min} U_{bh}/h$		
Case	F_x'	F_y'	F_z'	F_x'	F_y'	F_z'	F_x'	F_y'	F_z'
F10	0.24	0.25	0.32	0.24	0.30	0.29	0.45	1.10	0.89
F50	0.24	0.24	0.36	0.15	0.17	0.29	0.54	0.57	0.87

Table 3.9: Time scales of force fluctuations on a particle in case F10 and F50 normalised by h/U_{bh}. Considered time scales are the micro time scale, τ_m, the integral time scale, τ_ℓ, and the time-lag related to the first local minimum $\tau_{\min}$.

pronounced local minima of negative value. The profiles of F_x', F_y' and F_z' in case F10 exhibit local minima, however they are not as pronounced as in case F50. In particular, for the range of $\tau U_{bh}/h$ shown in figure 3.26, the correlations of F_x' and F_y' remain positive. For F_z' the correlations become negative, however the magnitude is close to zero.

Table 3.9 provides the values of τ_m, τ_ℓ and $\tau_{\min}$ as defined in §3.5.1 for the auto-correlations in figure 3.26. As in the smooth wall case (cf. table 3.6), the time scales are of order one when scaled in outer flow units. The micro time scales, $\tau_m U_{bh}/h$, are found to be of similar value in both cases for drag and lift fluctuations (cf. table 3.9). In case F10 (F50) the value of $\tau_m U_{bh}/h$ of F_z' is 33%

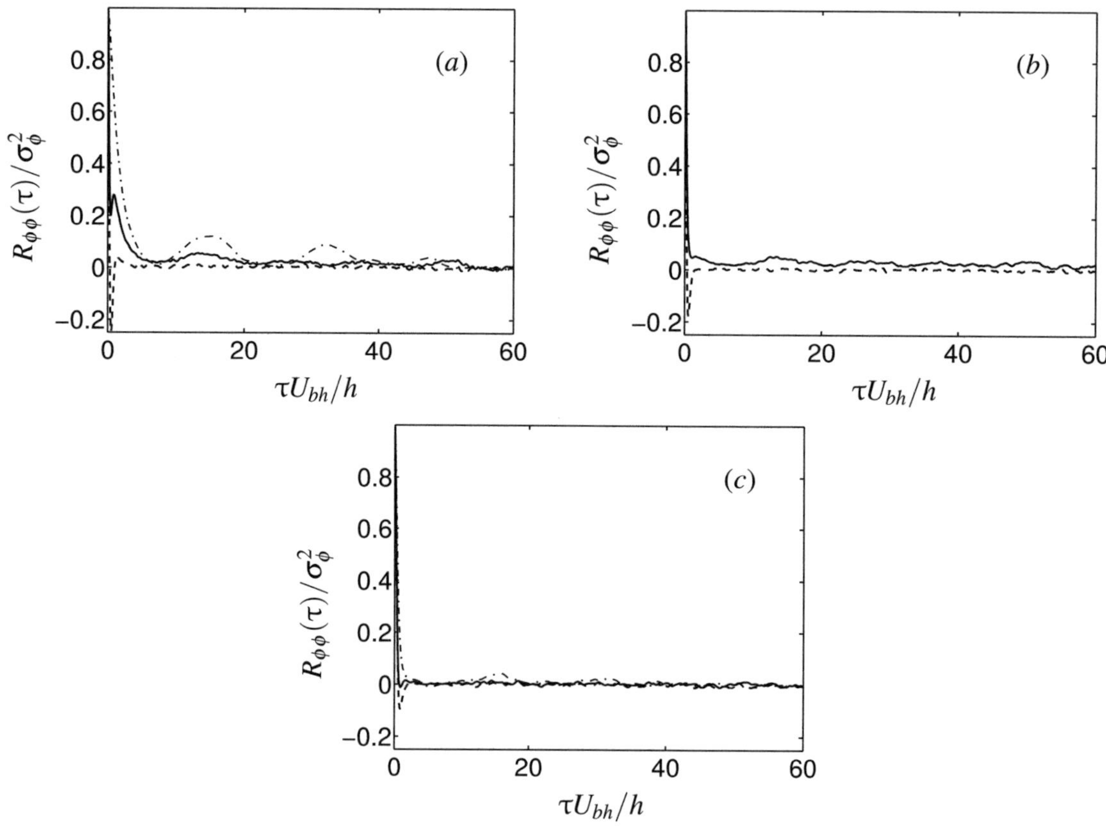

Figure 3.27: As figure 3.26 but for $0 \leq \tau U_{bh}/h \leq 60$.

(50%) larger than the respective value of F_x' (cf. table 3.9). These trends differ from the smooth wall results, (table 3.6), where $\tau_m U_{bh}/h$ of the force fluctuations increases with increasing s^+. Furthermore, $\tau_m U_{bh}/h$ of $\mathscr{F}_x'$ was found to be larger than $\tau_m U_{bh}/h$ of $\mathscr{F}_z'$. The differences could be related to the different shape of the auto-correlation, which appears to be strongly dominated by the occurrence of a local minimum for the force on a particle. The latter might be linked to the effect of pressure on the force on a particle. Recall, that the auto-correlation of pressure on a smooth wall in a turbulent channel flow is known to have pronounced minima of negative value similar to those in case F50 (see Kim, 1989 for auto-correlation in space, Quadrio & Luchini, 2003 for auto-correlation in time). However, it is remarkable that the effect of pressure on the force on a particle in case F10 should be that pronounced, as here the hydraulically smooth regime is approached for which it is generally assumed that viscous forces on the wall dominate the force behaviour.

It should be noted that, in contrast to the scaling by h/U_{bh}, τ_m increases for all fluctuation components from case F10 to case F50 when scaled by viscous units, i.e. by ν/u_τ^2 (table omitted).

Table 3.9 provides the time lag $\tau_{\min}$, related to the local minimum in the auto-correlation of figure 3.26. In case F10, the smallest value of $\tau_{\min}U_{bh}/h$ is obtained for F_x', the value increases by a factor of 2.4 (1.9) for F_y' (F_z'). The strong increase related to lift fluctuations might be heavily biased by a not well pronounced minimum (cf. figure 3.9*b*). In case F50, similar values of $\tau_{\min}U_{bh}/h$ are obtained for drag and lift fluctuation which increase by 61% for F_z' with respect to F_x'. In both cases the values of $\tau_{\min}U_{bh}/h$ for F_x' and F_z' are similar, in particular, in case F10 the values are respec-

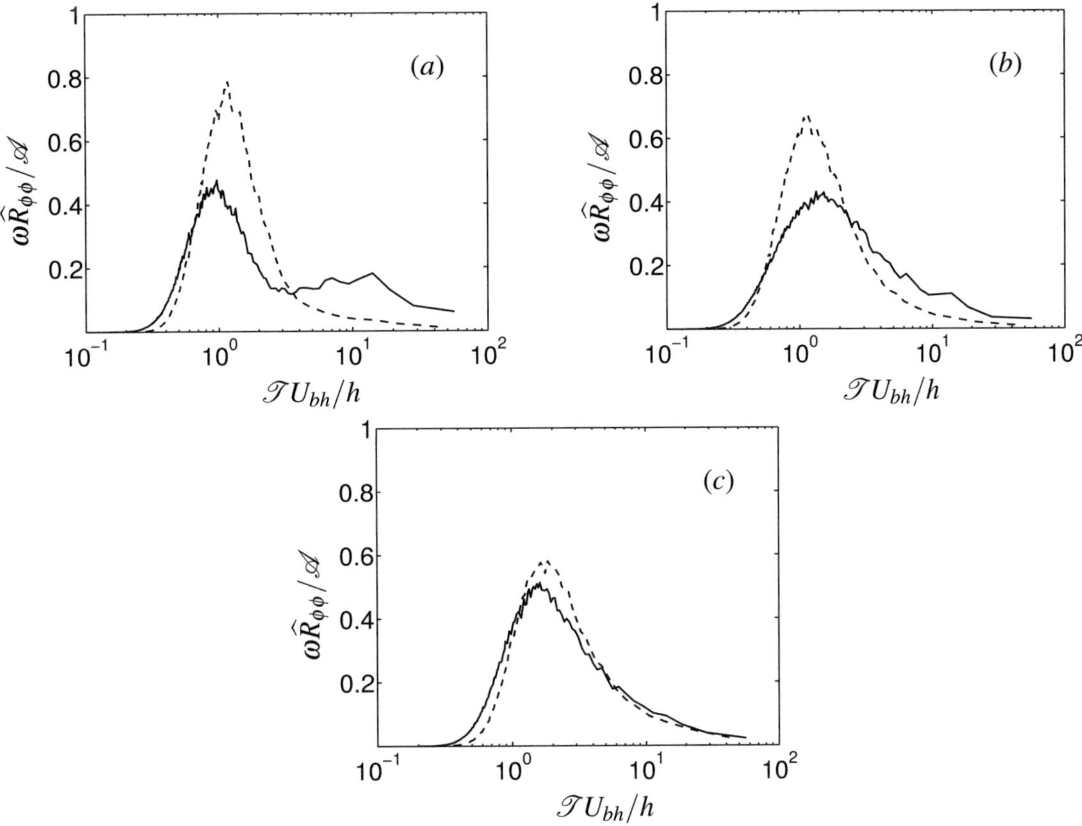

Figure 3.28: Pre-multiplied spectra $\omega\widehat{R}_{\phi\phi}$ of particle force fluctuations as a function of period $\mathscr{T}U_{bh}/h$. Spectra are normalised by $\mathscr{A}$ as defined in the text. Panels and line style as in figure 3.26.

tively 17% smaller and 2% larger as in case F50. It is interesting to note, that for the present cases $\tau_{\min}U_{bh}/h$ is 1.9 to 2.8 times larger than the value of $\tau_m U_{bh}/h$.

The integral time scale, $\tau_\ell U_{bh}/h$, is comparable to the micro time scale $\tau_m U_{bh}/h$ for all force fluctuation components (table 3.9), which indicates that the integral time scale as defined by (3.15) fails to provide a measure of large time scales in the present cases. A more appropriate definition based on the auto-correlations is difficult to define, as the characteristics of profiles in figure 3.26 vary among the cases.

The influence of the streamwise periodicity on the auto-correlations of particle force fluctuations appears to be weak in figure 3.27. Only in figure 3.27(*a*) some large scale fluctuations are noticeable in case F10. However, the amplitude is small and is damped out faster than in the smooth wall case. A possible explanation for the weaker influence might be the contribution of pressure on the force fluctuations. Recall, that pressure is commonly found to have a smaller correlation length than the shear stress at the wall (Kim, 1989; Choi & Moin, 1990; Quadrio & Luchini, 2003). Thus its influence might lead to a shortening of the correlation in time.

Figure 3.28 shows pre-multiplied spectra of the force fluctuations computed and normalised analogously to figure 3.23. Here, the comparison to the respective smooth wall forces is omitted. The figure reveals that most of the energy in the force fluctuation on a particle is contributed to time scales with periods of order h/U_{bh}. In particular the curves of $\omega\widehat{R}_{\phi\phi}/\mathscr{A}$ exhibit maxima with a period in the range of $\mathscr{T}U_{bh}/h = 1$ to 2 in both cases and for all three force components. The values of $\omega\widehat{R}_{\phi\phi}/\mathscr{A}$

$\mathscr{T}_{\max}U_{bh}/h$	F_x'	F_y'	F_z'
F10	1.0	1.3	1.6
F50	1.2	1.2	1.8

Table 3.10: Period of local maximum in pre-multiplied spectra of force fluctuations, $\mathscr{T}_{\max}$, normalised by outer flow time scale, U_{bh}/h.

in case F10 exceed the values of case F50 for small and large $\mathscr{T}U_{bh}/h$. In particular for F_x' an increased contribution of time scales with a large periods can be observed. This might be linked to an increased bias of the domain size in streamwise direction in this case F10. The larger values of $\omega\widehat{R}_{\phi\phi}/\mathscr{A}$ for small and large periods in case F10 are compensated by more distinct maxima in case F50. Interestingly, the maxima are most pronounced for F_x' and F_y' with values being nearly twice as large in case F50 than in F10. In contrast, the differences in the spectra of F_z' are smaller and can be found in particular for smaller periods.

The decrease of $\omega\widehat{R}_{\phi\phi}/\mathscr{A}$ with the particle size D^+ for small $\mathscr{T}U_{bh}/h$ is in agreement with the filtering argument utilised in the discussion of the auto-correlation of particle torque above. The decrease of $\omega\widehat{R}_{\phi\phi}/\mathscr{A}$ with the particle size D^+ for large $\mathscr{T}U_{bh}/h$ is more difficult to explain and might be a result of multiple factors. It might be related to the damping of large-scale flow structures in the near wall region by roughness as discussed in §3.7.2 (cf. figure 3.8). Also it could indicate an increased effect of pressure in case F50 which might lead to an amplification of smaller scales and reduce the relative influence of large scales structures.

The periods related to the maxima of the pre-multiplied spectra in figure 3.28, $\mathscr{T}_{\max}U_{bh}/h$, are given in table 3.10. Keep in mind, that $\mathscr{T}_{\max}$ represents a range of scales rather than a single pronounced peak (cf. figure 3.28). The table reveals, that the value of $\mathscr{T}_{\max}U_{bh}/h$ for drag fluctuations is 1.0 in case F10 and shifts to the somewhat higher value of 1.2 in case F50. The maximum in the pre-multiplied spectra of lift fluctuations is positioned at a period of $\mathscr{T}_{\max}U_{bh}/h = 1.3$ (1.2) in case F10 (F50) which is 30% (0%) higher than for drag. The value $\mathscr{T}_{\max}U_{bh}/h$ of the spanwise force fluctuation is 60% (50%) larger than for drag fluctuation in case F10 (F50). Comparing the values of $\mathscr{T}_{\max}U_{bh}/h$ in table 3.10 with $\tau_{\min}U_{bh}/h$ in table 3.9 shows that $\mathscr{T}_{\max} \approx 2\tau_{\min}$ for most quantities.

Here, the most dominant scales $\mathscr{T}U_{bh}/h$ are found in the range of 0.4 to 4. This agrees with the study of Hofland (2005) (cf. figure 5.11, p. 90) and Hofland *et al.* (2005) (cf. figure 10). The authors estimated drag and lift by the difference of two pressure measurements on a cube with dimensions of about $3300\delta_v$ at Re_b of about $1.3{\cdot}10^5$. The ratio of open channel depth to cube height was about six, which is similar to the ratio of h/D in case F50. They found that the highest values in the pre-multiplied spectra of estimated drag and lift for periods in the range of $\mathscr{T}U_{bh}/h = 0.6$ to 6.

3.5.5 Time scales of cross-correlation between drag and lift fluctuations on a particle

In section 3.5.4 it is shown that in the present case the time scales of drag and lift fluctuations, i.e. $\tau_{\min}U_{bh}/h$ and $\tau_m U_{bh}/h$, are comparable with each other (cf. table 3.9). Also the spectra in figure 3.28 behave similarly for drag and lift fluctuations. In contrast to this the time scales as well as the decrease of the spectra for large periods from case F10 to F50 were found to differ for spanwise force fluctuations. This could be seen as an indication that, in the current setup, drag and lift fluctuations are related to similar flow structures and might be closely correlated in time, while spanwise force fluctuations are relate to somewhat different flow structures.

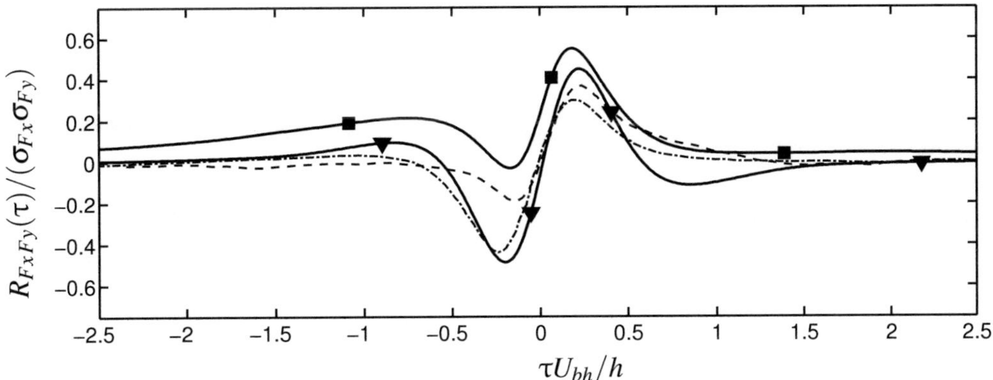

Figure 3.29: Cross-correlation of drag and lift fluctuation, $R_{FxFy}(\tau)/(\sigma_{Fx}\sigma_{Fy})$, as a function of time lag $\tau h/U_{bh}$. Figure shows results of case F10 (■) and case F50 (▼) in comparison with the indirect measurements of Hofland (2005), figure 6.5(a), lowest protrusion, for a cube with a height of $3300\delta_v$ at $\mathrm{Re}_b = 1.3 \cdot 10^5$, (– · –) and with the direct measurements of Dwivedi (2010), figure 7.23(a), zero protrusion, of spheres with $D^+ = 3100$ at $\mathrm{Re}_b = 1.68 \cdot 10^5$ (– – –).

Figure 3.29 presents the cross-correlation of drag and lift fluctuations defined by (3.14) with $\phi = F'_x$ and $\psi = F'_y$. For $\tau = 0$, the correlation of drag and lift fluctuations is small, i.e. 0.23 (−0.06) in case F10 (F50). However, a significant correlation occurs at a non-zero time lag. The cross-correlation reaches a maximum value of $R_{FxFy}(\tau)/(\sigma_{Fx}\sigma_{Fy}) = 0.550$ (0.450) for a time lag of $\tau U_{bh}/h = 0.19$ (0.24) in case F10 (F50). Thus, on average lift fluctuation follow drag fluctuation of equal sign with a time lag of about $0.2h/U_{bh}$. The cross-correlation reaches a local minimum of value $R_{FxFy}(\tau)/(\sigma_{Fx}\sigma_{Fy}) = -0.03$ (−0.49) for a time lag of $\tau U_{bh}/h = -0.17$ (−0.20) in case F10 (F50). While in case F10 the correlation is essentially zero at the local minimum, in case F50 it reaches negative values, indicating that drag and lift fluctuations are on average of opposite sign at a time lag of $\tau U_{bh}/h = -0.20$. As the time lag reaches higher negative values the correlation becomes positive once more, such that zero correlation is approached from above as the time lag goes to $-\infty$.

In addition to the results of case F10 and F50, the results of two experimental studies are presented in figure 3.29. One data set was taken from Hofland (2005), figure 6.5(a) at his lowest protrusion height. The author indirectly measured drag and lift by the difference of two pressure measurements on a cube with a height of $3300\delta_v$ at $\mathrm{Re}_b = 1.3 \cdot 10^5$. The ratio of channel width to open channel height was 3.0 and the ratio of cube height to open channel height 5.6. The other data set is taken from Dwivedi (2010), figure 7.23(a), who directly measured drag and lift on a spherical particle with $D^+ = 3100$ in a hexagonal packing of spheres with zero protrusion at $\mathrm{Re}_b = 1.68 \cdot 10^5$. The ratio of H/D was 5.4 and thus comparable to the one of case F50 and the measurements of Hofland (2005). The width of the open channel flow was $W/H = 2.1$ which is also comparable to the one of Hofland (2005). The latter ratio should be considered low and in both experiments strong influences of secondary currents have to be expected. In spite of the different Reynolds numbers and particle shapes considered in the experiments, the obtained cross-correlations agree well with the results of case F10 and F50. Largest differences are found in the value of the maximum and the minimum.

To study the comparison in more detail and address some scaling issues table 3.11 lists the difference between the time lag of maximum and minimum in figure 3.29, τ_Δ, normalised by the six time scales derived in §3.5.1. As can be seen, the value of τ_Δ normalised by a scaling which involves u_τ in-

Case	$\tau_\Delta U_{bh}/D$	$\tau_\Delta U_{bh}/h$	$\tau_\Delta u_\tau/D$	$\tau_\Delta u_\tau/h$	$\tau_\Delta u_\tau^2/\nu$	$\tau_\Delta(u_\tau U_{bh})/\nu$
F10	6.01	0.34	0.40	0.023	4.3	65
F50	2.01	0.42	0.16	0.034	7.9	98
DWI10	2.05	0.38	0.21	0.038	637	6373
HOF05	2.56	0.46	0.36	0.065	1199	8437

Table 3.11: Difference of time-lag related to maximum and minimum in cross-correlation of drag and lift, τ_Δ, in case F10 and F50. The difference is normalised by various time scales and compared to the value obtained from the results of Hofland (2005) (HOF05) and Dwivedi (2010) (DWI10)

U_c/U_{bh}	F'_x	F'_y	F'_z	T'_x	T'_y	T'_z
F10	0.71	0.67	0.68	0.63	0.56	0.62
F50	0.58	0.61	0.58	0.51	0.46	0.54

Table 3.12: Convection velocities, U_c, of force and torque fluctuations on particles in case F10 and F50 normalised by U_{bh}

U_c/u_τ	F'_x	F'_y	F'_z	T'_x	T'_y	T'_z
F10	10.8	10.2	10.4	9.6	8.5	9.5
F50	7.2	7.5	7.2	6.4	5.6	6.7

Table 3.13: Convection velocities, U_c, of force and torque fluctuations on particles in case F10 and F50 normalised by u_τ

creases with increasing D^+. Scaling τ_Δ using D/U_{bh} leads to strong difference between case F10 and the other results. The best agreement is found when τ_Δ is scaled by h/U_{bh}, i.e. $\tau_\Delta U_{bh}/h = 0.40 \pm 0.6$.

To conclude, it was found that drag and lift are related though with a shift in time. Profiles of drag and lift correlations from experiments at much higher Reynolds numbers are of comparable shape. As one of the experiments approximated the drag and lift by few pressure measurements only, the agreement with direct measurements of the forces and the present results point towards a strong influence of pressure on the relation. This interpretation is in favour of a model approach proposed by Hofland (2005) based on a convected frozen pressure field. Also, the agreement might be partly explained by a lack of scale separation. In the simulation this is due to the small Reynolds number considered, but also due to a small ratio of h/D. In the experiments the particle Reynolds numbers are large, and force fluctuation on the particle might be expected to be independent of the viscous scale (cf. §3.7.4). However, the ratio of open channel height to characteristic roughness length is small and thus h/U_{bh} might be the relevant time scale for these configurations.

3.6 Convection velocities of force and torque fluctuations

In turbulent flows, such as open channel flow, flow structures are convected downstream with a certain velocity. The convection velocity of the structures, U_c, is sometimes used to relate length scales of a turbulent field to time scales. To do so, it is assumed that the flow does not alter as it is convected downstream, which is commonly referred to as Taylor's frozen turbulence hypothesis (Taylor, 1938).

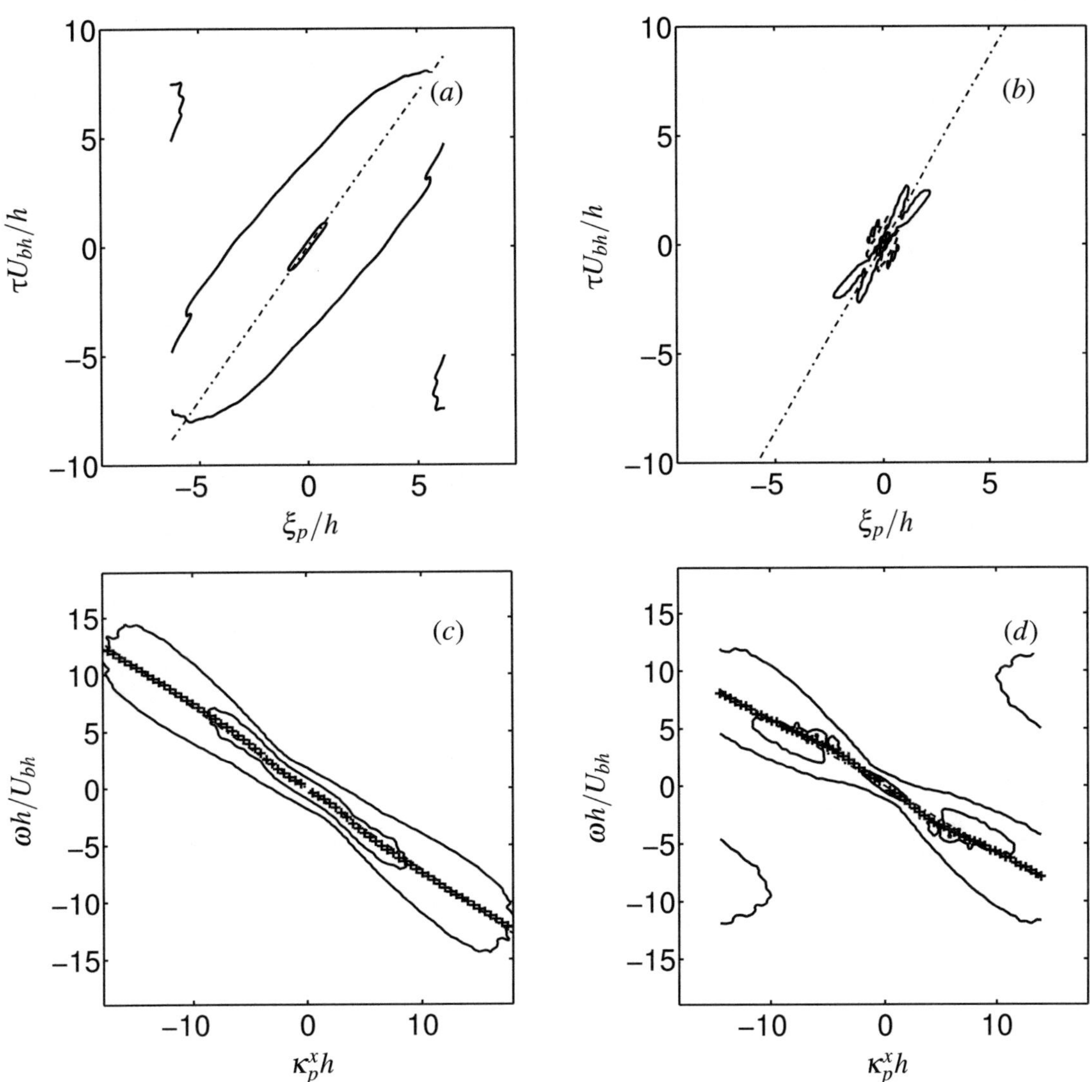

Figure 3.30: Space-time correlation of drag fluctuations in case F10 (*a*,*c*) and F50 (*b*,*d*). (*a*,*b*) Space-time correlation in physical space normalised by $(\sigma_x^F)^2$ as a function of space lag ξ_p/h and time lag $\tau U_{bh}/h$. Lines show iso-contours of [0.05 0.5] (———), and iso-contour at value [−0.05] (*b* only, – – –). (*c*,*d*) Space-time correlation in spectral space normalised by integral of spectral space over range of normalised wave numbers and normalised periods. The correlation is shown as a function of frequency, $\omega h/U_{bh}$, and wave number, $\kappa_p^x h$. Lines show iso-contours at values [0.001 0.01] (———). The symbols (+) mark $\omega_c(\kappa_p^x)$ as defined in the text. In all panels dash-dotted lines illustrate the global convection velocity U_c.

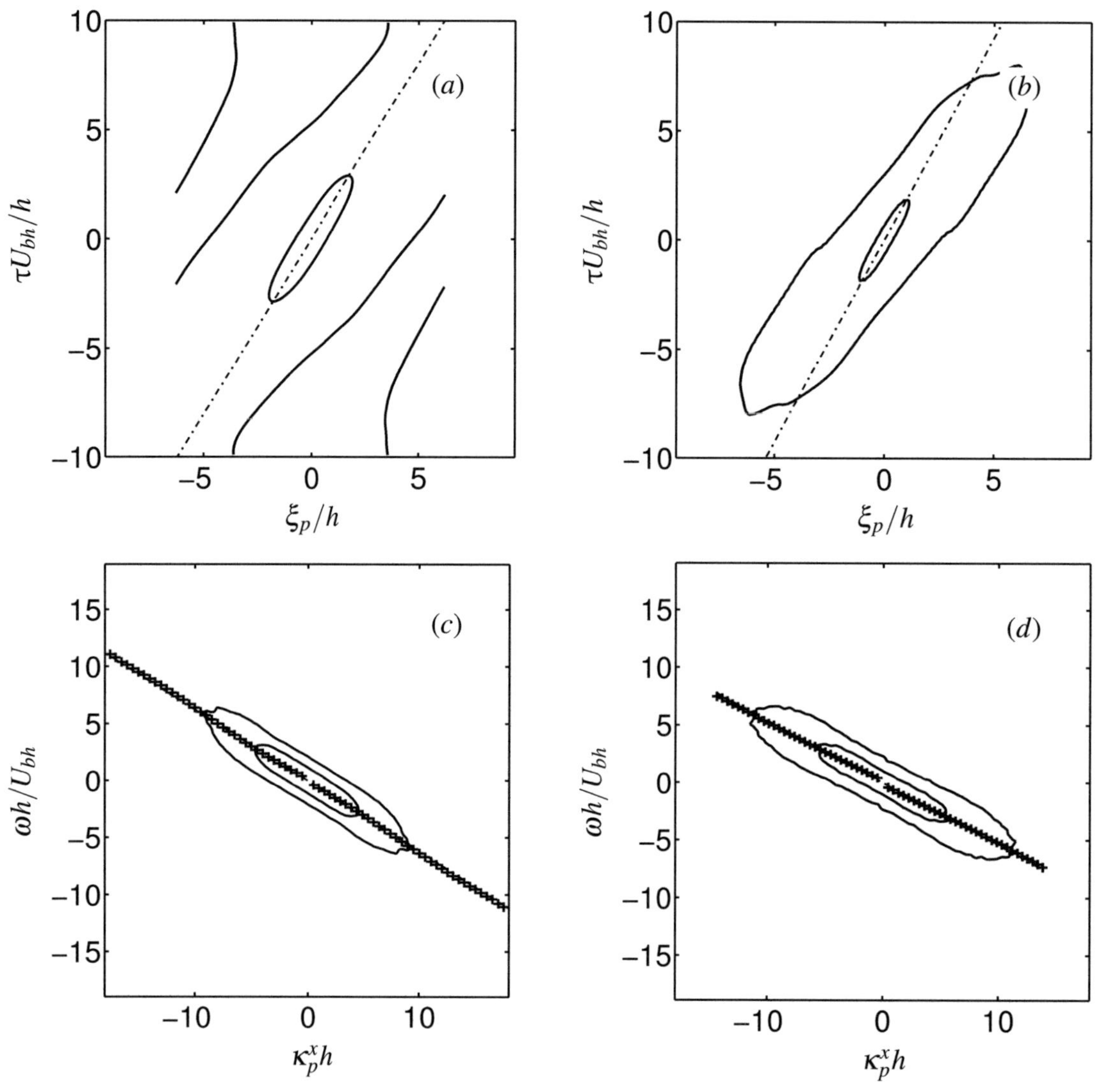

Figure 3.31: As in figure 3.30 but for spanwise torque fluctuations.

A detailed discussion on the issue can be found in del Álamo & Jiménez (2009) and the references therein.

This section focuses on the convection velocities of force and torque fluctuations as a result of turbulent flow structures. The definition of the convection velocity used in the present context requires a space-time correlation of force or torque fluctuations, which in physical space can be defined as

$$R_{\xi_p\tau}(\xi_p,\tau) = \frac{1}{t_1-t_0}\int_{t0}^{t1} \phi(x_p,y_p,z_p,t)\ \phi(x_p+\xi_p,y_p,z_p,t+\tau)\ \mathrm{d}t\,, \tag{3.17}$$

where $\phi(\mathbf{x}_p,t)$ is a component of the force or torque on the particle, τ is the time lag and $\xi_p = n\Delta_p^x$ with $n \in [-N_p^x/2-1, N_p^x/2]$ the streamwise coordinate of particles with respect to the particle centred at $\mathbf{x}_p = (x_p,y_p,z_p)$. Here, N_p^x is the (even) number of particles in streamwise direction and Δ_p^x the distance of particles in streamwise direction. Recall, that periodicity is applied in streamwise direction, i.e. $\xi_p = \xi_p + nL_x$. Also note, that ξ_p is a discrete quantity, which is stressed by the suffix p. To increase the quality of the correlation defined by (3.17) sample averaging over the number of particles N_p can be carried out. For reasons of efficiency the space-time correlations were not computed as above in physical space but in spectral space by employing the method of Welch (1967) in time (cf. §3.5 and a Fourier transform in spanwise direction.

Several methods to define convection velocities in turbulent flow have been proposed (see for example Quadrio & Luchini, 2003 or del Álamo & Jiménez, 2009). It should be noted, that a convection velocity of a flow field component cannot be defined unambiguously as scales with different lengths might travel at different speeds. The methods of defining convection velocities are often based on the space-time correlation of a signal, $R_{\xi\tau}(\xi,\tau)$, or its Fourier transform in spectral space, $\widehat{R}_{\kappa\omega}(\kappa^x,\omega)$, as function of wave number in streamwise direction, κ^x, and frequency, ω. Here, the convection velocity, U_c, of a particle quantity is defined as a weighted average of the convection velocity $u_c(\kappa_p^x)$ related to a given wave number, κ_p^x. del Álamo & Jiménez (2009) propose to define u_c by the centre of mass $\omega_c(\kappa_p^x)$ of the profile $\widehat{R}_{\kappa\omega}(\kappa_p^x,\omega)$ for the given κ_p^x, i.e. $u_c = -\omega_c(\kappa_p^x)/\kappa_p^x$. However, in the present case such a definition leads to a strong bias due to the aliasing of energy in the large wave numbers (small wave lengths) caused by the discrete nature of particle distribution which limits the smallest wavelength that can be resolved in streamwise direction to the distance of the particles. The aliasing results in the existence of contour lines at positive and negative frequencies for wavenumbers of large magnitude (cf. figure 3.30*b*,*d*). Therefore here, $\omega_c(\kappa_p^x)$ is defined by the centre of mass of the one-sided frequency spectrum, i.e. only the range $-\,\mathrm{sgn}(\kappa_p^x)\omega_c > 0$ is considered. This leads to a bias for very small wave numbers but reduces the otherwise dominant bias from the large wave numbers. From the values of $u_c(\kappa_p^x)$ a global convection velocity, U_c, is defined by the weighted average

$$U_c = -\frac{\int_{\kappa_p}\int_{\omega} u_c\,|\widehat{R}_{\kappa\omega}(\kappa_p^x,\omega)|\,(\kappa_p^x)^2\,\mathrm{d}\omega\mathrm{d}\kappa_p}{\int_{\kappa_p}\int_{\omega}|\widehat{R}_{\kappa\omega}(\kappa_p^x,\omega)|\,(\kappa_p^x)^2\,\mathrm{d}\omega\mathrm{d}\kappa_p}. \tag{3.18}$$

This average weights the convection velocities of each wave number with the energy contained in the respective wave number. Thus the higher energetic wave numbers contribute more to the global convection velocity defined by (3.18) (cf. del Álamo & Jiménez, 2009). The above definitions become clearer when studying figure 3.30 and figure 3.31 that show the space-time correlation of drag and spanwise torque fluctuation in physical and spectral space. The symbols $+$ in figure 3.30(c,d) and figure 3.31(c,d) show $\omega_c(\kappa_p^x)$ as defined above. The convection velocities U_c are illustrated in form of straight dash dotted lines.

Table 3.12 and table 3.13 list the convection velocities of force and torque fluctuation normalised by U_{bh} and u_τ respectively. In case F10 the value of U_c/U_{bh} range from 0.46 (T_y') to 0.71 (F_x'). These values are in the range which is commonly reported in smooth wall flows. For example, for smooth wall channel flow del Álamo & Jiménez (2009) reported values of U_c/U_{bh} in the range of 0.5 to 0.8 and 0.6 to 0.8 for the streamwise and spanwise velocity fluctuations close to the wall, respectively.

In terms of normalisation by u_τ the value of U_c/u_τ in case F10 are in the range of 8.5 (T_y') to 10.8 (F_x'). This generally agrees well but for wall-normal torque is somewhat lower than the values commonly reported in smooth wall flow. In smooth wall channel flow as well as boundary layers, the near wall velocity fluctuations are convected at a speed larger than $9u_\tau$ (Quadrio & Luchini, 2003; del Álamo & Jiménez, 2009; Krogstad *et al.*, 1998). The convection velocity of shear stress fluctuations and pressure fluctuation on a smooth wall in channel flow are studied by Jeon *et al.* (1999). The authors report values of 9.6 (10.4) for the streamwise (spanwise) shear stress and 13.1 for the pressure fluctuations. However, Jeon *et al.* (1999) point out, that the convection velocity of scales at a given frequency vary from these averaged values. The values of U_c/u_τ found for case F10 are within this range of variation.

The somewhat lower values of U_c/u_τ in case F10 compared to smooth wall results from the literature might also be linked to a weak effect of roughness. The convection velocity used to apply the Taylor hypothesis is often lower in rough wall cases than in the smooth wall cases. For example Krogstad & Antonia (1994) used a 18% smaller value for the convection velocity in the rough wall case based on the free stream velocity. Table 3.12 and table 3.13 show that the convection velocity of force as well as torque fluctuations decrease from case F10 to case F50. In particular, U_c/U_{bh} (U_c/u_τ) in case F50 is on average 16% (31%) smaller than in case F10. The values of U_c/U_{bh} in case F50 range from 0.46 (T_y') to 0.58 (F_y') corresponding to a range of values of U_c/u_τ of 5.6 (T_y') to 7.5 (F_y'). These convection velocities are in the range of convection velocities given for the wall-normal vorticity provided by Flores & Jiménez (2006). The authors found that the convection velocity very close to a wall with velocity disturbances differs from the one of a smooth wall and is of similar value as the local average streamwise velocity ($\pm 2u_\tau$ within $0 < y^+ < 100$) and in the range $U_c/u_\tau = 1.5$ to $U_c/u_\tau = 7$.

3.7 Flow structures related to force and torque fluctuations

This section studies the relation of flow structures to force and torque fluctuations using correlation functions between the flow field and force and torque fluctuations. Three configurations are considered, (i) the correlation of flow structures in a smooth wall flow to force on a smooth wall surface element, (ii) the correlation of flow structures to torque on particles, (iii) the correlation of flow structures to force on particles. Here, the analysis is limited to streamwise and spanwise force and torque fluctuations for which only some correlations will be discussed in the following. The relation of flow structures to force on a smooth wall element will be analysed in §3.7.1, the relation of flow structures to force and torque on spherical particles will be analysed in §3.7.2. A simplified view on the relation of flow structures to force and torque will be reconsidered in section §3.7.3. Some scaling aspects will be considered in §3.7.4.

3.7.1 Flow structures related to force and torque fluctuations on a surface element in smooth wall flow

In the following, the relation of flow structures to force fluctuation on a square surface element in smooth wall channel flow is studied by means of correlation functions. The analysis is limited to the streamwise and spanwise force component as defined by (3.12). The correlation function between the force on a square surface element and the flow field is defined as

$$R_{\phi\psi}(\xi,y,\zeta) = \frac{1}{t_1 - t_0} \int_{t0}^{t1} \phi(x_s+\xi, y, z_s+\zeta, t)\ \psi(x_s, 0, z_s, t)\ \mathrm{d}t\,, \tag{3.19}$$

where $R_{\phi\psi}(\xi,y,\zeta)$ is the correlation function between the scalar quantities ϕ and ψ, $\phi(\mathbf{x},t)$ is a component of the flow field, $\psi(\mathbf{x}_s,t)$ is a component of the force on the square element centred at $\mathbf{x}_s = (x_s, 0, z_s)$, ξ and ζ are the spatial shift in streamwise and spanwise direction with respect to the element's centre and t_0 and t_1 define a time interval. To increase the quality of the correlation defined by (3.19), sample-averaging over the number of elements is carried out. In practice, this can be evaluated efficiently by employing a fast Fourier transform in the periodic directions. The correlation function of the fluctuations is obtained by subtracting the plane-average from the correlation function.

The results are based on 72 snapshots from a smooth wall simulation at $\mathrm{Re}_\tau = 183$ in an interval of $300H/U_{bH}$ using a pseudo-spectral method (cf. appendix B for details). The symmetry of the obtained correlations with respect to the ξ-axis can serve as a measure of statistical convergence. This appears to be reasonable in all cases.

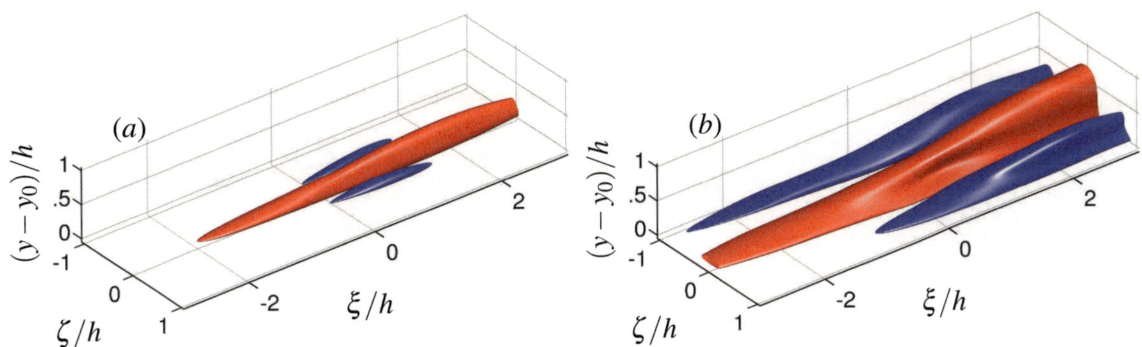

Figure 3.32: Iso-surfaces of cross-correlation, $R(\xi, y, \zeta)/|R|_{\max}$, of the streamwise force fluctuation on a square surface element, $\mathscr{F}_x'$, and streamwise velocity fluctuation, u' at values 0.15 (red) and -0.15 (blue). Panels show results for two different side length of the square element, (*a*) $s^+ = 11$ and (*b*) $s^+ = 46$.

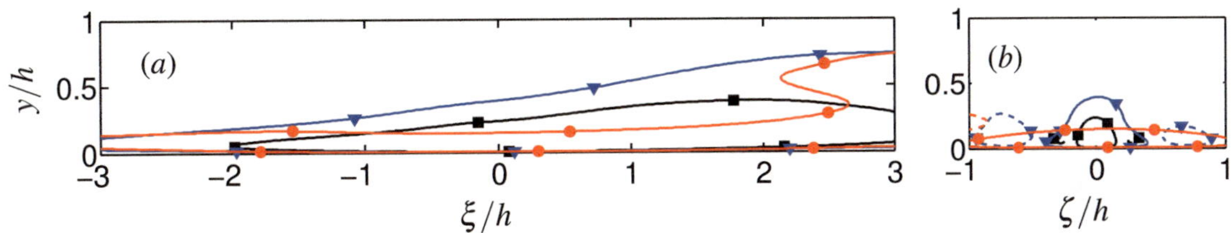

Figure 3.33: Cross-correlation, $R(\xi, y, \zeta)/|R|_{\max}$, of the streamwise force fluctuation $\mathscr{F}_x'$ on a square smooth wall surface element with side length s and streamwise velocity fluctuation u'. (*a*) Plane at zero spanwise shift, $\zeta = 0$, as function of y/h and streamwise shift ξ/h, (*b*) plane at zero streamwise shift, $\xi = 0$, as function y/h and spanwise shift ζ/h. Solid (dashed) lines show 0.15 (-0.15) of the maximal amplitude in each case. Colours and symbols indicate results for a square element with side length $s^+ = 11$ (■), $s^+ = 46$ (▼), $s^+ = 138$ (•).

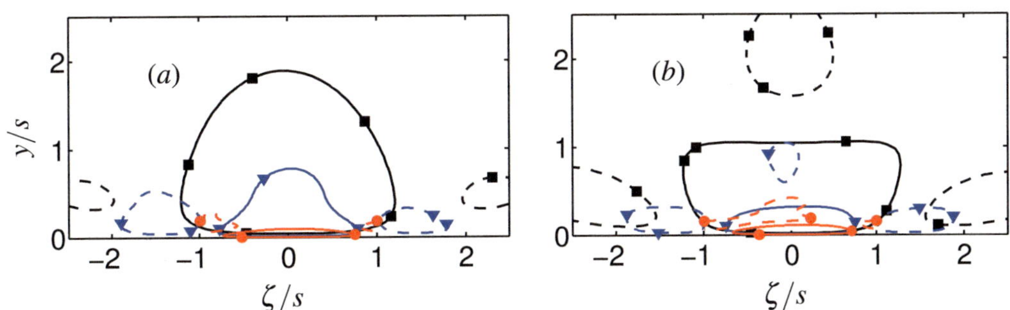

Figure 3.34: (*a*)As figure 3.33(*b*) but axis scaled by s. (*b*) as figure 3.36(*b*) but axis scaled by s.

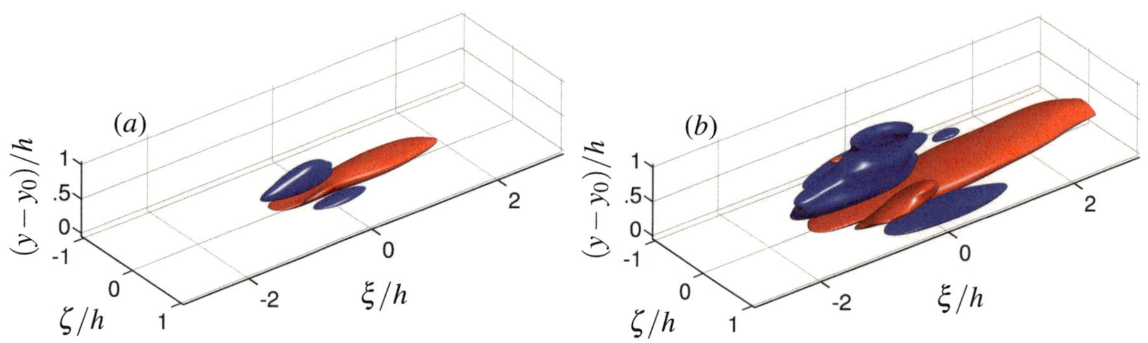

Figure 3.35: Iso-surfaces of cross-correlation, $R(\xi,y,\zeta)/|R|_{\max}$, of the spanwise force fluctuation on a square surface element, $\mathscr{F}_z'$, and spanwise velocity fluctuation, w' at values 0.15 (red) and -0.15 (blue). Panels show results for two different side length of the square element, (*a*) $s^+ = 11$ and (*b*) $s^+ = 46$.

Figure 3.32 shows the three-dimensional correlation function of streamwise velocity fluctuations u' with the fluctuations of drag on a square element, $\mathscr{F}_x'$, for two different side lengths $s^+ = 11$ and $s^+ = 46$. The correlation function is visualised by iso-surfaces of fractions of the maximal correlation $|R|_{\max}$, i.e. at values of $R(\xi,y,\zeta)/|R|_{\max} = 0.15$ (-0.15) in red (blue). Note, that $|R|_{\max}$ is non-zero for the correlations studied below.

Cross-sections at zero spanwise and streamwise separation are shown in figure 3.33 as contour lines of values $R(\xi,y,\zeta)/|R|_{\max} = 0.15$ (-0.15) with continuous (dashed) lines. Here, additionally the results for a square smooth wall surface element with side length $s^+ = 138$ is shown. Note that this corresponds in the present case to a side length of about $3/4h$ which has to be considered as large.

The figures reveal long streamwise elongated structures of positive correlation in the vicinity of the squared element positioned at the origin $(\xi,y,\zeta) = (0,0,0))$. This is in line with the idea that positive (negative) velocity fluctuations, u', are related to positive (negative) drag fluctuations, $\mathscr{F}_x'$, on a smooth wall surface element. The regions of significant positive correlation are large and expand over $5h$ ($s^+ = 12$) to $9h$ ($s^+ = 46$ in streamwise direction (cf. figure 3.33*a*) and are thus much larger than s and in particular for $s^+ = 46$ and $s^+ = 138$ of length comparable to the domain length. In wall-normal direction the region of significant correlation reaches heights larger than $0.4h$, indicating that in the present case the outer flow is of relevance for drag on the surface element.

In addition to the regions of positive correlation figure 3.32 and figure 3.33 reveal the existence of regions with significant negative correlations which flank the positive correlation at a spanwise distance. The alternating area of positive and negative correlations might be related to high speed and low speed streaks which have a spacing of about 50 to 70 wall units in smooth wall flows. However, in particular figure 3.33(*b*) shows that the spanwise distance of the regions of negative correlation increase with s^+. Also, figure 3.34(*a*) indicates, that the spanwise extent of the correlation might scale with s for large values of s^+ at higher Reynolds numbers. Here, the positive correlation above the element expands over approximately $2s$ for all side lengths s declining somewhat as s^+ increases. This could seen as an indication that as s increases systems of alternating streamwise velocity fluctuations become relevant for drag fluctuations which have spanwise and wall-normal extensions larger than the commonly reported system of near wall structures. It should be noted that the contours of $s^+ = 138$ do not follow all trends, however this might be due to the very large side length comparable to the domain height. In particular, for $s^+ = 138$ the correlation expands over the entire domain.

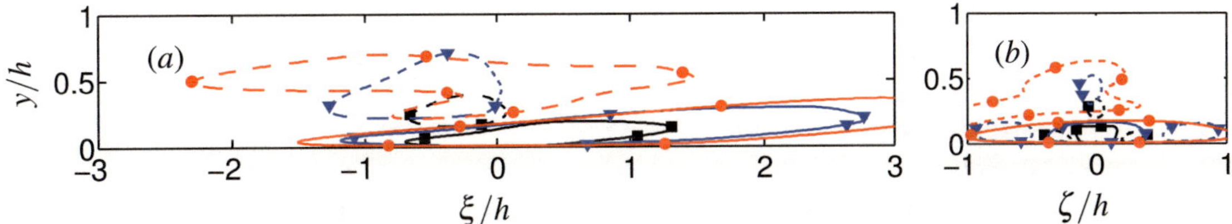

Figure 3.36: Cross-correlation, $R(\xi, y, \zeta)/|R|_{\max}$, of the spanwise force fluctuation $\mathscr{F}_z'$ on a square smooth wall surface element with side length s and spanwise velocity fluctuation w'. (*a*) Plane at zero spanwise shift, $\zeta = 0$, as function of y/h and streamwise shift ξ/h, (*b*) plane at zero streamwise shift, $\xi = 0$, as function y/h and spanwise shift ζ/h. Solid (dashed) lines show 0.15 (−0.15) of the maximal amplitude in each case. Symbols indicate results for a square element with side length $s^+ = 11$ (■), $s^+ = 46$ (▼), $s^+ = 138$ (•).

Figure 3.35 and figure 3.36 display the correlation of the spanwise force fluctuation, $\mathscr{F}_z'$, with the spanwise velocity fluctuations, w'. Once more a positive correlation is found in the vicinity of the element centre, which is in line with the idea that positive (negative) velocity fluctuations in spanwise direction, w', lead to positive (negative) spanwise force on a square smooth wall surface element, $\mathscr{F}_z'$. In comparison with the iso-surfaces of figure 3.32 the iso-surfaces in figure 3.35 have a more complex shape. The streamwise extent of the region of significant positive correlation increases from less than $2h$ for $s^+ = 11$ to more than $4h$ for $s^+ = 134$ (figure 3.36*a*). Interestingly the increase is most pronounced in the downstream direction. Similar to before, regions of significant negative correlation exist, here however at various locations and with complex shapes (cf. figure 3.35). Figure 3.36(*a*) can be used to quantify the regions of significant correlation visible away from the wall upstream of the centre of the wall elements. It shows that as s^+ increases, the negative correlation moves further upstream and further away from the wall reaching up to $y/h = 0.5$. Figure 3.34(*b*) displays the correlation at zero streamwise separation with axis scaled by s. It is found, that the region of significant positive correlation in the vicinity of the wall element expand over $2s$ in spanwise direction, decreasing slightly as s increases. Again this indicates that some properties of the correlation scale with s.

One can conclude that the character of the correlation of $\mathscr{F}_z'$ and w' is very different to the correlation of $\mathscr{F}_x'$ and u' (cf. figure 3.33*a*). Still some similarities might be recognised, i.e. both correlations contain streamwise elongated structures and regions of alternating signs in spanwise direction which decrease from a spanwise length of $2s$ for $s^+ = 11$ to approach s as s^+ increases.

Generally, the regions of significant correlation between flow structures and streamwise and spanwise force fluctuation on a smooth wall element are of sizes of the order of h. The streamwise extent of the regions were found to increase from $s^+ = 11$ to 46. For $s^+ = 138$ the spanwise extent of the correlations are of order of the domain size. The results follow the idea that positive (negative) velocity fluctuations are related to positive (negative) force fluctuations on the smooth wall surface element directed in the same coordinate direction. However, the regions of significant correlation exhibit an additional complexity that goes beyond this simplified consideration which reflect the complexity of turbulent flow and the near-wall system of flow structures. It might be interesting to carry out a similar analysis with the flow fields at higher Reynolds numbers available from the simulations of del Álamo & Jiménez (2003) and Hoyas & Jiménez (2008). Also some more insight could be gained from investigating additional correlations. However, the objective of this section is to provide a reference for the following discussions for which the presented analysis is sufficient.

	F10	F50	ξ^{F10}_{max}/D	y^{F10}_{max}/D	ζ^{F10}_{max}/D	ξ^{F50}_{max}/D	y^{F50}_{max}/D	ζ^{F50}_{max}/D
$F'_x\,;u'$	1.50	0.48	6.29	1.86	−0.07	−1.13	1.11	−0.02
$T'_z\,;u'$	2.19	1.53	4.86	1.86	−0.07	0.54	1.11	−0.02
$F'_z\,;w'$	0.50	0.43	−1.29	1.57	0.00	−0.15	1.09	0.00
$T'_x\,;w'$	0.66	0.71	1.00	1.64	0.00	0.46	1.11	0.00
$F'_x\,;p'$	0.83	1.26	−2.57	0.86	−0.07	−0.74	0.98	−0.02
$F'_z\,;p'$	0.98	1.09	0.79	2.21	1.64	0.11	1.09	−0.83

Table 3.14: Maximal amplitude of correlation of particle force and torque fluctuations with flow field fluctuations, $|R|_{max}$, in case F10 and case F50. $|R|_{max}$ is normalised by u_τ ($\rho_f u_\tau^2$) as a characteristic measure of the velocity (pressure) fluctuation and the standard deviation of the force/torque fluctuation. Additionally the position of $|R|_{max}$ with respect to ξ/D, y/D and ζ/D are provided.

3.7.2 Flow structures related to force and torque on a particle

In the following, the relation of flow structures to force and torque fluctuation on a particle is studied by means of correlation functions. The correlation function between the force and torque on a particle and the flow field is defined as

$$R_{\phi\psi}(\xi,y,\zeta) = \frac{1}{t_1 - t_0}\int_{t0}^{t1} \phi(x_p+\xi,y,z_p+\zeta,t)\;\psi(x_p,y_p,z_p,t)\;\mathrm{d}t\,, \tag{3.20}$$

where $R_{\phi\psi}(\xi,y,\zeta)$ is the correlation function between the scalar quantities ϕ and ψ, $\phi(\mathbf{x},t)$ is a component of the flow field, $\psi(\mathbf{x}_p,t)$ is a component of the force or torque on the particle with particle centre positioned at $\mathbf{x}_p = (x_p,y_p,z_p)$, ξ and ζ are the spatial shift in streamwise and spanwise direction with respect to $\mathbf{x}_p$ and t_0 and t_1 define a time interval. The quality of the correlation (3.20) is increased by sample-averaging over the number of particles N_p. The cross-correlation function of the fluctuations is evaluated by subtracting the plane average from the correlation.

In total 24 possible correlations exists, of which only some are presented below. The correlations have been computed from the flow and particle fields collected during run-time of the simulations case F10 and case F50 (cf. appendix §C.3 for more details). The correlations are presented below in form of iso-surfaces and contours at ± 0.1 or ± 0.15 of their respective maximal magnitude, $|R|_{max}$. It is not explicitly pointed out in the following when the thresholds do not lead to iso-contours/surfaces of both, positive and negative levels. The values of $|R|_{max}$ of the considered correlations are given in table 3.14. Here, $|R|_{max}$ is normalised by u_τ and $\rho_f u_\tau^2$ as a characteristic scale for the velocity and pressure fluctuations, respectively, and the standard deviation of the respective force or torque fluctuation. Table 3.14 shows that all values are non-zero and of order one. Once more, the statistical convergence of a correlation can be judged by deviations from the symmetry (or asymmetry) with respect to the ξ-axis.

Figures 3.37 to 3.39 present the cross-correlation of drag fluctuation on a particle, F'_x, with the streamwise velocity fluctuations, u', in case F10 and case F50. Figure 3.37 illustrates the correlations in form of iso-surfaces at $\pm 0.15\,|R|_{max}$ in three-dimensions scaled by h. Additionally, figure 3.38 and figure 3.39 display the correlation in form of iso-contours of value $\pm 0.15\,|R|_{max}$ in cross-sections of zero streamwise and spanwise separation. In figure 3.38 the axes are normalised by h, in figure 3.39 the axes are normalised by D.

All figures exhibit positive values of the correlation in the vicinity of the particle centre. The regions of significant correlation are of similar size when scaled with h (cf. figure 3.37 and figure 3.38). In streamwise direction regions of significant correlation expand over about $6h$ ($4.5h$) in case F10

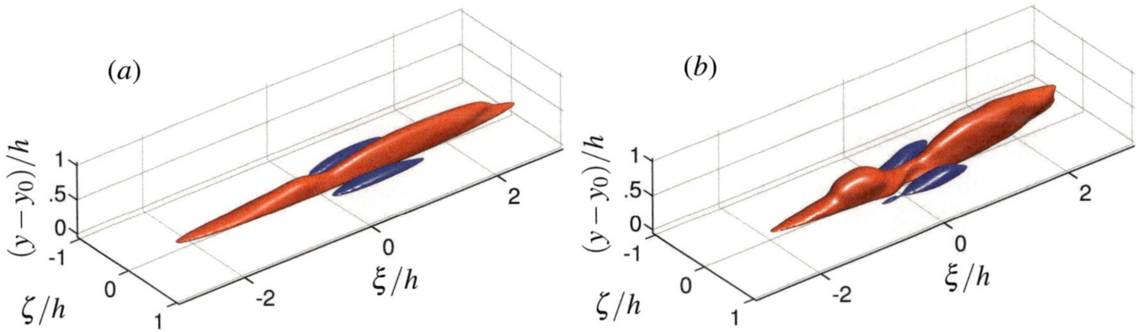

Figure 3.37: Iso-surfaces of cross-correlation, $R(\xi,y,\zeta)/|R|_{\mathrm{max}}$, of the streamwise particle force fluctuation, F'_x, and streamwise velocity fluctuation, u' at values 0.15 (red) and -0.15 (blue). Panels show case F10 (*a*) and case F50 (*b*).

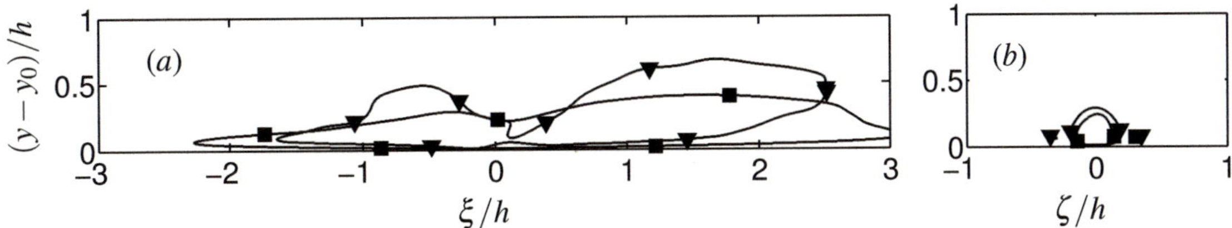

Figure 3.38: Cross-correlation, $R(\xi,y,\zeta)/|R|_{\mathrm{max}}$, of the streamwise particle force fluctuation, F'_x, and streamwise velocity fluctuation, u'. (*a*) Plane at zero spanwise shift, $\zeta = 0$, as function of $(y-y_0)/h$ and streamwise shift ξ/h; (*b*) plane at zero streamwise shift, $\xi = 0$, as function $(y-y_0)/h$ and spanwise shift ζ/h. Solid (dashed) lines show 0.15 (-0.15) of the maximal amplitude in each case. Symbols indicate results for case F10 (■) and F50 (▼).

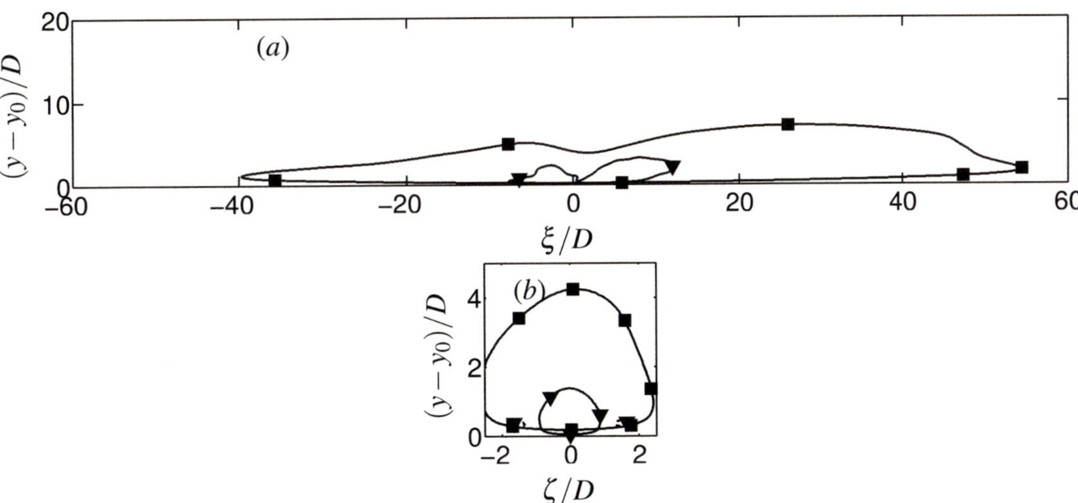

Figure 3.39: As figure 3.38 but axes scaled with particle diameter D.

(F50) (cf. figure 3.38*a*), which corresponds to about $100D$ ($22D$) (cf. figure 3.39*a*). The wall-normal distances in which significant correlation is obtained differ. Highest values of $(y-y_0)/h$ are reached in case F50 (figure 3.38*a*) for which the region of significant correlation appears to be somewhat lifted away from the wall. But also in case F10, the region expands over a substantial fraction of h, indicating a large influence of h. The spanwise size of the region of positive correlation in figure 3.38(*b*) is similar in both cases and is about $0.25h$. This corresponds to about $4.5D$ ($2D$) in case F10 (F50) in figure 3.39(*b*).

The uplifted shape of the correlation function in case F50 is intriguing. A possible explanation could be the effect of roughness on the large scales. As can be seen in figure 3.8 and has been argued by Flores & Jiménez (2006) roughness appears to shorten the large scales of streamwise velocity in particular for $(y-y_0)/h < 0.4$. The damping of the large structures in this region could correspondingly shorten the correlation in particular close to the wall and cause the lack of correlation for $\xi/h > 1.5h$ and $(y-y_0)/h < 0.4$ in case F50 (cf. figure 3.38*a*) in contrast to case F10.

Figure 3.37 shows that in both cases regions of considerable negative correlation appear. The shape of the iso-surfaces are similar, the one of case F10 being slightly more elongated. In both cases the correlation intersects little with the plane at $\zeta = 0$, which leads to the small contours at $-0.15\,|R|_{\max}$ observed in figure 3.38(*b*). The region of significant negative correlation has a size of about $h \times 0.25h \times 0.25h$ in streamwise, wall-normal and spanwise direction respectively. The centre of the region is located downstream of the particle position at a spanwise distance of about $\zeta/h = \pm 0.3$. This corresponds to about 55 (70) wall units in case F10 (F50) and compares well to the distance of high speed streak and low speed streak commonly found in smooth wall flows (cf. Smith & Metzler, 1983; Kim *et al.*, 1987).

The results of the correlation of F_x' and u' discussed above are in the following compared to the results of the correlation of $\mathscr{F}_x'$ and u' in figure 3.32 and figure 3.33. It is found, that the contour lines shown in case F10 in figure 3.38 agree rather well with the contour lines of figure 3.33 for $s^+ = 11$. Thus here, the flow structures related to drag on a particle appear to be comparable to those related to drag on a smooth wall. This motivates a more detailed comparison between the two correlations despite fundamental differences of drag on a particle and drag on a smooth wall surface element. The good agreement in particular for case F10 could be seen as an indication that for spheres small enough with respect to the the smallest scales of flow motion, the correlation between particle force and flow field fluctuations approaches the correlation of near-wall stress and flow field fluctuations.

In contrast to the good agreement between the smooth wall results and the results in case F10, the agreement of the results of case F50 in 3.37 and 3.38 to the results in figure 3.32 and figure 3.33 for $s^+ = 46$ is less good. Firstly, the correlation close to the wall shortens from case F10 to case F50, while on the contrary the correlation in the smooth wall simulation increases from $s^+ = 11$ to $s^+ = 46$ (cf. figure 3.33*a*, figure 3.32). Secondly, the iso-surfaces in figure 3.37(*b*) and the the contour lines in figure 3.38(*a*) show a pronounced drop between $-0.5 < \xi/h < 0.5$. A similar drop, albeit at a much smaller extent, is also present in case F10. In contrast to this, the respective results in the smooth wall case, that is the iso-surfaces in figure 3.32 as well as the contour lines in figure 3.33 do not exhibit a pronounced drop. The above implies that the fundamental difference between drag on a sphere and on a smooth wall element becomes more pronounced with increasing sphere size. As discussed above, the shortening of the correlation in case F50 might be linked to the effect of roughness on the flow scales not present in the smooth wall simulation. One could speculate that the drop found in the correlation might be related to the effect of pressure on drag fluctuations. This issue will be addressed further below.

Figures 3.40 and 3.41 show the cross-correlation between spanwise torque fluctuations on the particle, T_z', and streamwise velocity fluctuations, u'. From figure 3.40 it can be observed that the

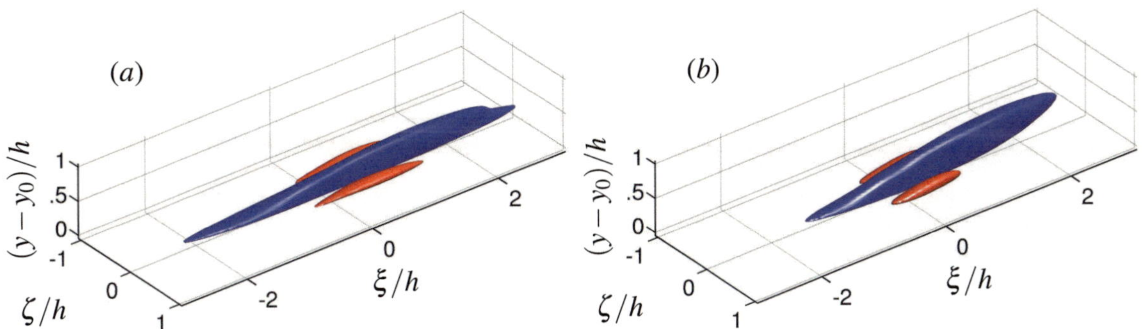

Figure 3.40: Iso-surfaces of cross-correlation, $R(\xi, y, \zeta)/|R|_{max}$, of the spanwise particle torque fluctuation, T_z', and streamwise velocity fluctuation, u' at values 0.15 (red) and −0.15 (blue). Panels show case F10 (*a*) and case F50 (*b*).

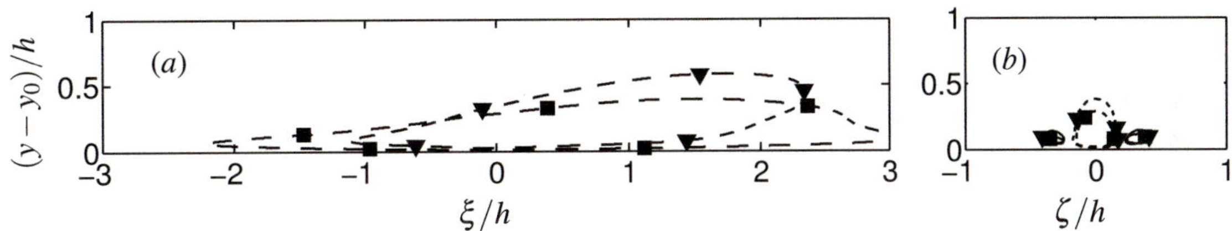

Figure 3.41: Cross-correlation, $R(\xi, y, \zeta)/|R|_{max}$, of the spanwise particle torque fluctuation, T_z' and streamwise velocity fluctuation u'. Panels and lines as in figure 3.38.

correlation exhibits similarities to the correlation of F_x' and u' shown in figure 3.37 albeit at opposite sign. That is in both cases, regions of significant negative negative correlation are located in the vicinity of the sphere. The shown iso-surfaces are elongated in streamwise direction and of sizes much larger than the particle diameter. This finding is in agreement with the idea that structures of positive (negative) u' relate on average to negative (positive) spanwise torque fluctuations on a particle. Similar to before, the regions significant negative correlation are flanked by smaller but likewise elongated regions of significant correlation with alternating sing. Also similar to before, the identified region in figure 3.41(*a*) shortens in its streamwise extend from case F10 to case F50 and the appears to be more inclined to the ξ-axis in case F50. A quantitative comparison of the correlation between spanwise torque on a particle and u' (figure 3.41) with the correlation of drag on a particle and u' discussed with figure 3.38 leads to good agreement but also reveals some clear differences. In particular, the characteristic drop discussed for the correlation of F_x' and u' is not present in the correlation of T_z' and u'. Recalling that torque on a spherical particle is only due to viscous forces this results supports the idea that the drop in the correlation of F_x' and u' might be related to the effect of pressure on drag.

Figure 3.42 shows the correlation of drag fluctuations, F_x', with pressure fluctuations, p', in case F10 and F50 illustrated by iso-surfaces of the value $\pm 0.1\,|R|_{max}$. Figure 3.43(*a*) shows the corresponding iso-contours at zero spanwise separation, i.e. $\zeta = 0$. In both cases, drag fluctuations correlate with equally signed pressure fluctuations upstream of the particle and pressure fluctuations of opposite sign downstream of the particle. The shape and size of the region of significant positive correlation is sim-

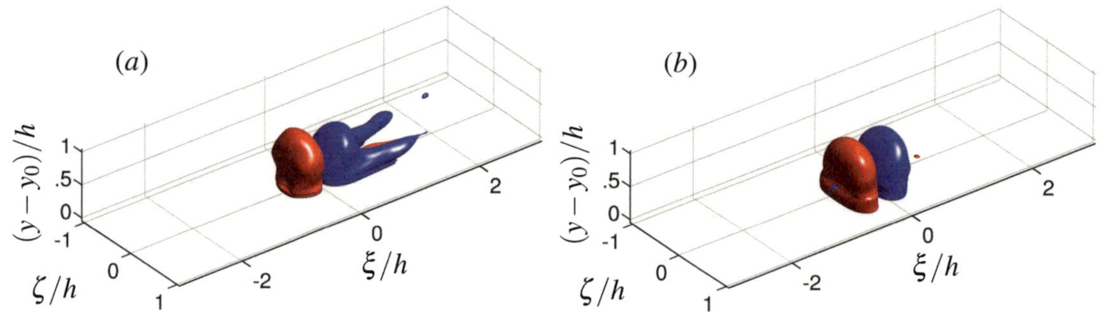

Figure 3.42: Iso-surfaces of cross-correlation, $R(\xi, y, \zeta)/|R|_{max}$, of the streamwise particle force fluctuation, F_x', and pressure fluctuation, p' at values 0.1 (red) and -0.1 (blue). Panels show case F10 (*a*) and case F50 (*b*).

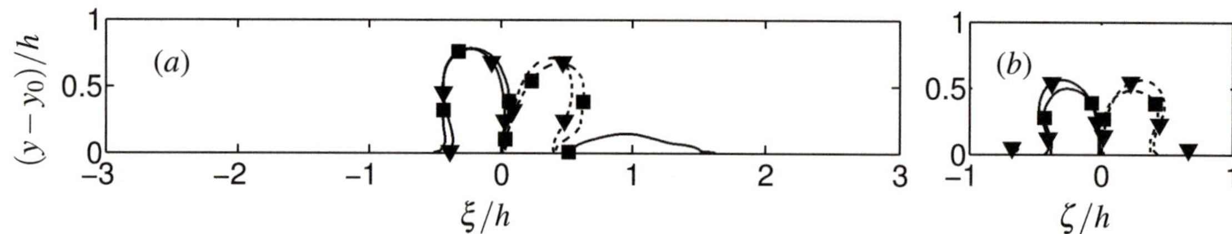

Figure 3.43: (*a*) Cross-correlation, $R(\xi, y, \zeta)/|R|_{max}$, of the streamwise particle force fluctuation, F_x', and pressure fluctuation, p', at zero spanwise shift, $\zeta = 0$. (*b*) Cross-correlation, $R(\xi, y, \zeta)/|R|_{max}$, of the spanwise particle force fluctuation, F_z', and pressure fluctuation, p', at zero spanwise shift, $\xi = 0$. Solid (dashed) lines show 0.1 (-0.1) of the maximal amplitude in each case. Symbols indicate results of case F10 (■), and case F50 (▼).

ilar in both cases, i.e. it is $0.5h \times 0.7h \times h$ in streamwise, wall-normal and spanwise direction. The shape of iso-surfaces at $-0.1\,|R|_{max}$ matches in size in both cases but exhibits an additional fork-type extensions in case F10 (figure 3.42). Furthermore, in case F10 an region of positive correlation is located close to the wall at $0.5 < \xi/h < 1.5$ (cf. figure 3.43*a*). Note, that in contrast to pressure regions in the three-dimensional time-averaged pressure field with similar high and low values upstream and downstream of the particle (cf. figure 3.10) the dimensions of the present iso-surfaces are larger than the particle diameter D, e.g. $10D$ ($3D$) in wall-normal direction for case F10 (F50).

The location of the correlation of F_x' and p' presented in figure 3.42 and figure 3.43 coincides with the location of the drop in the correlation of F_x' and u' (e.g. figure 3.38). This points once more towards the role of pressure on the drag fluctuation and the related correlations.

Figure 3.45 and figure 3.44 provide the correlation of spanwise particle force fluctuation, F_z', with spanwise velocity fluctuations, w'. Regions of significant positive correlation are observed in the vicinity of the particle, which is in line with the idea that F_z' correlates on average with equally signed structures of w' in the vicinity of the particle. Regions of negative values of the correlation occur next to those of positive correlation in both spanwise directions, and additionally for $-0.9 < \xi/h < -0.2$ and $0.2 < \zeta/h < 0.5$. The iso-surfaces in figure 3.45 are of similar size and similar, when scaled with h and of rather complex shape. The iso-surfaces extend about $1.5h \times 0.5h \times 1h$ in streamwise, wall-normal and spanwise direction.

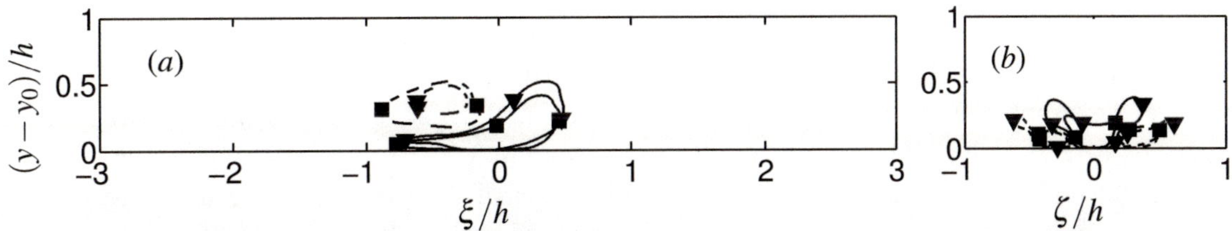

Figure 3.44: Cross-correlation, $R(\xi,y,\zeta)/\,|R|_{\max}$, of the spanwise particle force fluctuation, F_z', and streamwise velocity fluctuation, w'. (*a*) plane at zero spanwise shift, $\zeta = 0$, as function of $(y-y_0)/h$ and streamwise shift ξ/h; (*b*) plane at zero streamwise shift, $\xi = 0$, as function $(y-y_0)/h$ and spanwise shift ζ/h. Solid (dashed) lines show 0.1 (−0.1) of the maximal amplitude in each case. Symbols indicate results for case F10 (■), and case F50 (▼).

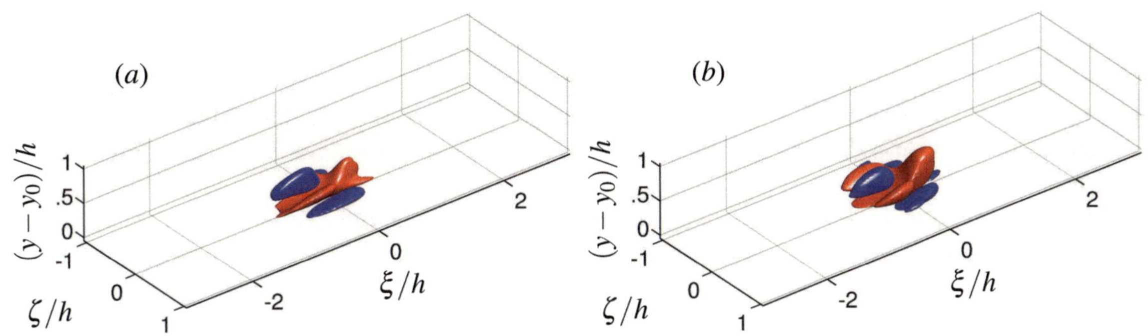

Figure 3.45: Iso-surfaces of cross-correlation, $R(\xi,y,\zeta)/\,|R|_{\max}$, of the spanwise particle force fluctuation, F_z', and spanwise velocity fluctuation, w' at values 0.1 (red) and −0.1 (blue). Panels show case F10 (*a*) and case F50 (*b*).

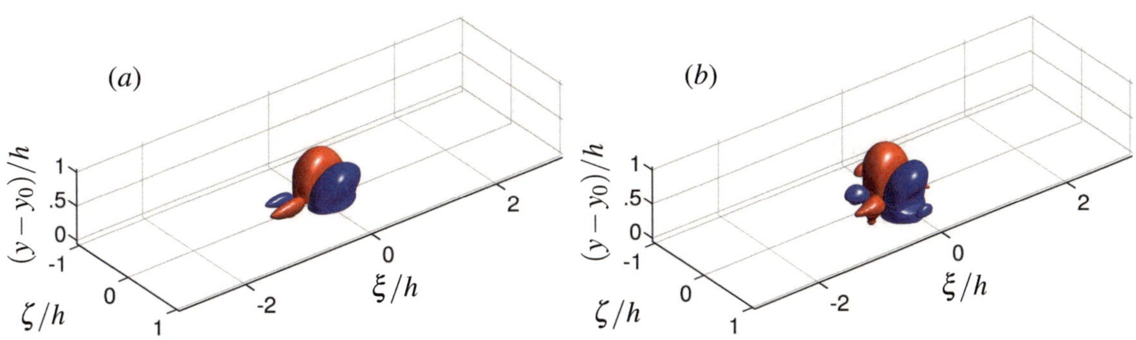

Figure 3.46: Iso-surfaces of cross-correlation, $R(\xi,y,\zeta)/\,|R|_{\max}$, of the spanwise particle force fluctuation, F_z', and pressure fluctuation, p' at values 0.1 (red) and −0.1 (blue). Panels show case F10 (*a*) and case F50 (*b*).

The correlation of F_z' and w' might be compared to the respective correlation of $\mathscr{F}_z$ and w' in the smooth wall results of figure 3.35 and figure 3.36. Some characteristic are found to be similar from the correlation of spanwise force on a particle to w' and spanwise force on a smooth wall surface element to w'. For example the existence of regions of positive and negative correlation in figure 3.36(a) and figure 3.44(a), as well as the contours of alternating sign in 3.36(b) and 3.44(b). However, differences are evident and more pronounced as in the comparison of drag on a particle and a wall to u'. In particular, in the region of significant correlation on a smooth wall in figure 3.35 is large, elongated in streamwise direction which increases with s^+. In contrast, the regions of significant simulation in figure 3.45 is confined in streamwise direction and differs little from case F10 to case F50.

Figure 3.46 displays the cross-correlation of spanwise force fluctuations, F_z', with pressure fluctuations, p', by iso-surfaces at $\pm 0.1\,|R|_{\max}$. Figure 3.43(b) shows the corresponding contours of the correlation in a cross-section at zero streamwise separation $\xi = 0$. The correlation of F_z' with p' in the vicinity of the particle is positive for $\zeta < 0$ and negative for $\zeta > 0$ which agrees with the idea that a negative (positive) pressure gradient in spanwise direction will cause a positive (negative) spanwise force on the particle. This simple picture is complicated by two smaller iso-surface upstream of the particles with alternating sign of their downstream neighbours (cf. figure 3.46), as well as by an additional pair of smaller sized structures with equal sign, centred around $(y - y_0)/h = 0.5$ and $\xi/h = \pm 1$ (cf. figure 3.43b and figure 3.46). It is interesting to note, that the size and value of the correlation related to pressure and drag or spanwise force fluctuations are comparable in both simulations (cf. figure 3.43, table 3.14).

Figure 3.47 and figure 3.48 show the correlation function between the streamwise torque fluctuations, T_x', and the streamwise velocity fluctuations, w', at $\pm 0.1\,|R|_{\max}$. The regions of significant correlation in figure 3.47 are shown in form of an iso-surface of positive sign, flanked by two iso-surfaces of negative sign in spanwise direction, and one iso-surfaces of negative sign aligned along the ξ-axis at a larger distance to the wall centred upstream of the particle. Note, that a region of positive correlation is in agreement with the idea that positive (negative) structures of w' relate to positive (negative) streamwise torque fluctuations (cf. figure 3.12). Comparing the correlation of streamwise torque and w' (cf. figure 3.47 to the correlation of spanwise force and w' (cf. figure 3.45) similar characteristics but also differences of the complex structure can be identified. In contrast to the correlation of spanwise force and w' the correlation of streamwise torque and w' appears to be more elongated in streamwise direction and less angled to the ξ-axis (cf. figure 3.44 and figure 3.48). Once more, one can speculate that the difference between the correlations is due to the contribution of pressure on the spanwise force fluctuations, which does not contribute to streamwise torque fluctuations. Some support for this hypothesis could be gained from the comparison of figure 3.47 to the smooth wall result the correlation of $\mathscr{F}_z'$ and w' shown in figure 3.35. In particular the results of case F10 compare well with the results for $s^+ = 11$.

3.7.3 Considerations on the relation of flow structures to force and torque fluctuations

This section aims to re-consider some simplified arguments mentioned in the previous sections on the relation of structures of flow field fluctuations on force and torque fluctuations on a smooth wall surface element and particles. There it was found that streamwise velocity fluctuations, u', in the vicinity of the wall are related to equally oriented drag fluctuations on a smooth wall surface element, $\mathscr{F}_x'$, or on a particle, F_x', as well as to oppositely signed spanwise particle torque fluctuations, $-T_z'$. Analogously, it was found that spanwise velocity fluctuations, w', in the vicinity of the wall are related to equally oriented spanwise force fluctuations on a smooth wall surface element, $\mathscr{F}_z'$, or on a particle, F_z', as well as to equally signed streamwise particle torque fluctuations, T_x'. In case of a smooth wall

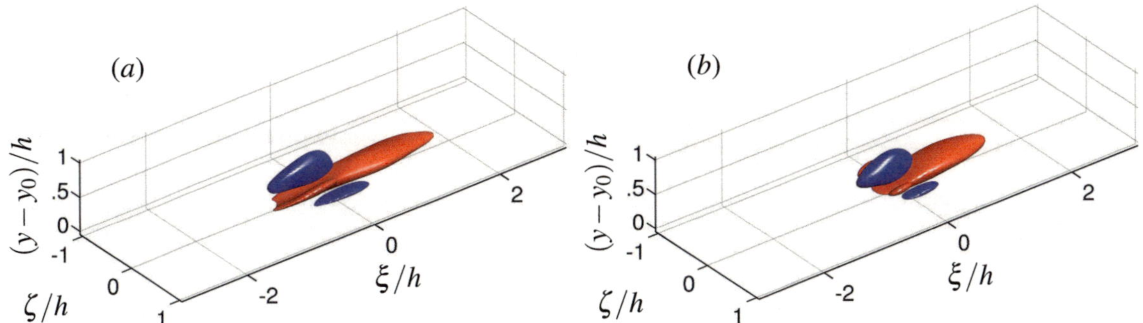

Figure 3.47: Iso-surfaces of cross-correlation, $R(\xi, y, \zeta)/|R|_{\max}$, of the streamwise particle torque fluctuation, T_x', and spanwise velocity fluctuation, w' at values 0.1 (red) and -0.1 (blue). Panels show case F10 (a) and case F50 (b).

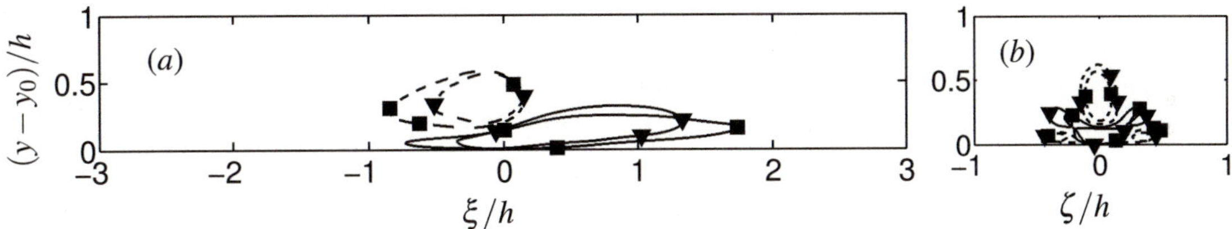

Figure 3.48: Cross-correlation, $R(\xi, y, \zeta)/|R|_{\max}$, of the streamwise particle torque fluctuation, T_x', and streamwise velocity fluctuation w'. Panels and lines as in figure 3.44.

this finding can be explained by (to a first order approximation) direct relation of the velocities close to the wall to the viscous stresses on the smooth wall, i.e. the shear stress at the wall reduces to

$$\boldsymbol{\tau} \cdot \mathbf{n} - p^{tot}\mathbf{n} = \begin{pmatrix} 2\mu\, \partial u/\partial y \\ 2\mu\, (\partial v/\partial y) - p^{tot} \\ 2\mu\, \partial w/\partial y \end{pmatrix}, \quad \text{for } \mathbf{n} = \begin{pmatrix} 0 \\ 1 \\ 0 \end{pmatrix}. \tag{3.21}$$

Additional considerations taking into account the integral character of force on a smooth wall element lead to the idea that only scales of lengths similar or larger than the smooth wall element will contribute to force fluctuations. In case of particles the normal vector to the surface is not constant which considerably complicates the aspect of stress at the wall and force or torque on the particle. Despite this, some results related to force and torque on a particle compare well with analogous results of the smooth wall. In particular results related to torque on particles, and especially in the case of small spheres with a limited roughness effect.

Torque on a spherical particle is only due to viscous stresses and might be largely caused by a force on a small surface area on the particle crest (cf. equation 3.10). Jointly with the weak effect of roughness on the flow structures in the present case this was seen as a possible explanation of the good agreement to the result of force on a smooth wall surface element. In contrast to torque, the force on a particle contains a contribution of viscous stresses and additionally a contribution of pressure (cf. equation 3.5). A negative pressure gradient across a particle in streamwise (spanwise) direction can be expected to result in a positive force contribution in streamwise (spanwise) direction. Note that here the gradient is important, e.g. an additional constant contribution to the pressure field around a particle will not influence the force on the particle. Hofland (2005), §6.2, pp. 94, discusses a modelled pressure distribution convected at a certain speed over spherical particles. From his studies the author concludes, that pressure fluctuations of size comparable to $1.5D$ are most efficient in contributing to pressure force fluctuations. It can be argued that pressure fluctuations of sizes much smaller and much larger than the particle diameter do not lead to significant pressure gradients across the particle thus cease influence.

The above consideration are consistent with the correlation functions studied in the previous sections. However, the correlations exhibit an additional complexity which can be seen as the footprint of coherent flow structures related to force and torque fluctuations.

3.7.4 Scaling aspects

Studying force and torque on particles by direct numerical simulation (DNS) is limited to low Reynolds numbers (cf. §2.2.2) and moderate ratios of h/D (cf. §3.1). As a result, the range of length scales of the turbulent flow is limited and a separation of scales is not expected. With only two simulations considered in the present study, conclusions with respect to scaling aspects are limited. Nevertheless, it is beneficial for the following discussion to consider some fundamental scaling concepts of flow structures related to force and torque fluctuations.

Assuming that the particle diameter is small compared to the effective open channel height, i.e. $D/h \ll 1$, two extreme cases with respect to the particle diameter can be considered: (i) $D^+ \ll 1$ and (ii) $D^+ \gg 1$. In case (i) the particle diameter is small compared to the viscous length. The smallest scales of turbulent motion will be much larger than the particle diameter. It seems reasonable to assume that in the limit of an infinitesimal small D, the flow structures related to force and torque on a particle might be comparable to the flow structures related with wall shear in smooth wall channel flow. Several studies on the latter flow structures can be found (Kravchenko *et al.*, 1993; Jeon *et al.*, 1999; Grosse & Schröder, 2009; Sheng *et al.*, 2009, among others). They show that (to a first order approximation) the flow structures related to wall shear scale in viscous units, i.e. with δ_v. Thus for case (i) in which $D/h \ll 1$ and $D^+ \ll 1$, the flow structures related to force and torque fluctuations on a particle might be expected to be independent of the particle diameter and to scale in viscous units.

In case (ii) the particle diameter is considered to be much larger than the viscous scale, $D^+ \gg 1$, and thus is much larger than the smallest scales of turbulent motion. In this case, the presence of the particle might have a strong influence on the flow itself, e.g. flow can separate behind spheres and lead to shedding of vortices with lengths comparable to D. As discussed in section 3.7.3, in case (ii) the length scale D might pose a limit for the size of flow structures that efficiently contribute to force and torque fluctuations on a particle. Only flow structures equal or larger than the particle diameter will effectively contribute to force or torque fluctuations, while the effect of flow motions smaller than D might cancel out due to the integral in the definition of force and torque. A question of interest is, if for a certain parameter range, the size of flow structures related to force and torque solely depend on D and is essentially independent of δ_v and h. One could speculate that this might be true when the following two assumptions hold. The first assumption is, that the structures related to force and torque fluctuations on a particle with $1 \ll D^+$ do not depend on the viscous length scale. Some support for such a hypothesis could be gained from assuming that the dynamics of these large structures only depends on inviscid processes. The second assumption is that the dominant flow structures related to the force and torque scale with D and are independent of h when $h/D \gg 1$. The latter assumption is similar to the assumption made in the derivation of the law of the wall, where it is assumed that for a certain wall distance far away from the outer flow the flow scales do not depend on h. Note that although the law of the wall has been found to be a good approximation of the mean velocity profile, the underlying assumption has been challenged in recent years. Some evidence exists that especially larger flow structures depend in parts on outer units (cf. del Álamo *et al.*, 2004 and Marusic *et al.*, 2010 for a recent discussion.)

The two hypotheses can be used to estimate the parameter range for which flow structures related to force and torque on a particle might scale with D only. For the viscous scale to lose influence, it seems reasonable to assume that D needs to be larger than the scales found to be of relevance to shear stress in the smooth wall channel flow, i.e. D^+ should be of the order of 1000. Assuming that a ratio of $h/D > 50$ is needed to ensure a minimal influence of h, leads to a friction Reynolds number of order $\mathrm{Re}_\tau = \mathcal{O}(10^4)$. As the highest Reynolds number in DNS of smooth wall channel flow have so far reached Re_τ up to 2000 (Hoyas & Jiménez, 2006), it is unlikely that such high Reynolds will be feasible in DNS in the near future. In experiments it seems more likely that Reynolds numbers of that order can be reached. While studies on flow over spheres did not meet the suggested range of parameters (cf. table C.1) some recent studies over natural gravel have approached it (Detert *et al.*, 2010*a*). It should be noted, that values of $h/D > 50$ and $\mathrm{Re}_\tau > 10^4$ might be commonly reached in rivers. For example, the river Rhine at Maxau has an average depth of about $h = 5$m, a bulk velocity of $U_{bh} = 1$m/s and a viscosity of water of about $\nu = 1 \cdot 10^{-6}$m^2/s leading to $\mathrm{Re} = 5 \cdot 10^6$ and Re_τ of the order of 10^5. In this case the criteria of $h/D > 50$ would hold for gravel with a characteristic size of up to 10cm which corresponds to an order of 1000 wall units.

In the present simulations, the Reynolds numbers and thus the range of scales in the flow are small. Also, the ratio of D/h might be considered moderately small for case F10 but it is large for case F50. The value of D^+ shows that in both cases the smallest scales of turbulent motion are comparable to the particle diameter. Thus, for the current setup the flow structures related to force and torque fluctuation on a particle might be influenced by δ_ν, D as well as by h.

3.8 Summary, conclusion and recommendation for future work

Direct numerical simulation of open channel flow over a geometrically rough wall was performed at a bulk Reynolds number of $\mathrm{Re}_b \approx 2900$. The wall consisted of a layer of spheres in a square arrangement touching a solid wall. Two particle diameters were considered: case F10 with $D^+ = 10.7$ ($\mathrm{Re}_\tau = 188$), and case F50 with $D^+ = 49.3$ ($\mathrm{Re}_\tau = 235$). In case F10 the effect of roughness on the flow field statistics was small, and the limit of the hydraulically smooth flow regime is approached. In case F50 the roughness effect was stronger, and the flow is in the transitionally rough flow regime.

The complexity of the time-averaged three-dimensional flow field within the roughness layer was discussed in detail. In both cases a re-circulation forms downstream of the spheres that is connected over the entire spanwise direction, being more pronounced in the large-sphere case. Three-dimensionality above the roughness layer is lost rapidly with wall distance, yielding a time-averaged flow field which is essentially one-dimensional beyond a distance of two particle diameters.

A main result of this chapter is the characterisation of the force and torque acting on a particle due to the turbulent flow. It was found that the mean drag on a sphere is 4% (case F10) and 15% (case F50) higher than the reference force $F_R = \rho_f u_\tau^2 A_R$, where A_R is the wall-normal projected area of the wall per particle. Given the definition of the friction velocity, these numbers reflect the fact that the drag force on the roughened bottom wall (below the fixed spheres) is small. In both cases a strong positive lift was obtained in agreement with previous experiments, exceeding values of 18% and 32% for F10 and F50, respectively, of the corresponding drag. The values of the mean spanwise torque on a particle are comparable to $-F_R r_R$, where r_R is the distance from the particle centre to the position of the virtual wall, located at a distance of $y_0 = 0.8D$ from the plane part of the solid wall. It was shown that in both cases the mean drag, lift and spanwise torque are to a large extent produced in a region of the particle surface which is located above the virtual wall ($y > y_0$). The spatial distribution over the particle surface of the stresses that lead to time-averaged drag, lift and spanwise torque are found to be similar in shape in the two cases.

It was observed that the intensity of particle force fluctuations (when normalised by F_R) is significantly larger in the large-sphere case. Conversely, when analysing the torque it is found that only the fluctuation intensity of the spanwise component is larger in the large-sphere case, whereas the two remaining components exhibit smaller fluctuation intensities when the sphere is larger. By means of a simplified model it was argued that the torque fluctuations might be explained by the spheres acting as a filter with respect to the size of the flow scales which can effectively generate torque fluctuations. As a model the shear forces and torque exerted by the flow on a square wall element in a smooth wall configuration were considered. By systematically varying the linear dimension of the wall element the influence of the length scale was analysed. Here it is found that the normalised fluctuation intensity of the streamwise and spanwise shear forces monotonically decreases with the filter size, while the wall-normal torque experiences a maximum of normalised fluctuation intensity for intermediate filter sizes of approximately 70 wall units. By assuming that the largest part of the particle torque fluctuations is generated in a small area around the particle tops, the results from the simplified model carry over to the corresponding components of the particle torque. Indeed a reasonable agreement was obtained between standard deviation and kurtosis of shear forces and torque acting on a square wall element on the one hand and respective particle torque components on the other hand. However, since only two particle sizes were considered, it cannot be stated with certainty that wall-normal particle torque fluctuations are indeed most intense at the above mentioned scale of 70 wall units. Similarly, based on the current data it is not possible to judge whether the model is capable of providing insight into the fully rough flow regime. These points should be clarified in future studies.

Fluctuations of both force and torque were found to exhibit strongly non-Gaussian probability density functions with particularly long tails. The deviation from a Gaussian distribution was significantly smaller in the large-sphere case, which was attributed to the smaller effect that highly intermittent small scales have on the larger-particle surface area. Moreover, it was observed that the spanwise torque component has a marked negative skewness. In the light of the analogy with a wall element in a smooth wall configuration, this finding is consistent with the positive skewness of the streamwise velocity fluctuations near the wall in a smooth wall channel flow.

The time scales of force and torque fluctuations were studied by means of auto-correlation functions. In the present case a separation of small and large time scales does not exist, as a result of the small Reynolds number considered. The time scales are of order one when scaled by the bulk velocity and effective channel height. An exception is the time scale of spanwise torque on the particle which appears to be biased by the streamwise domain size. In the large sphere case, the auto-correlation of force on the particle reveal pronounced minima with negative values. Local minima are also found in the small sphere case but of much lesser extent. The auto-correlation of torque on the particles differ little in both cases and do not exhibit a similar minima. It was speculated, that the minima are related to the pressure contribution to drag. Some support for this hypothesis comes from the auto-correlation of pressure at a smooth wall which is known to exhibit a pronounced minimum of negative value. The maximum of the pre-multiplied spectra reveals that the scale related to the minima in the auto-correlation contributes most to the force fluctuations on the particle.

The cross-correlation between drag and lift shows that the two particle quantities are closely related with a shift in time. The time scales of the correlation compare best when scaled by outer flow units. Comparison to experiments point towards a strong influence of pressure on the correlation, which is in favour of the model by Hofland (2005) that approximates the drag-lift correlation by results from a convected pressure field.

The convection velocity of force and torque on particles were studied with the aid of space-time correlations. In the small sphere case, the convection velocities are in the range commonly reported for smooth wall simulations when scaled with outer flow units ($U_c/U_{bh} = 0.56$ to 0.71), but somewhat lower than the values in the literature when scaled in viscous units ($U_c/u_\tau = 8.5$ to 10.8). In the large sphere case, the convection velocities are smaller ($U_c/U_{bh} = 0.46$ to 0.61, $U_c/u_\tau = 5.6$ to 7.5), which can possibly be linked to the effect of roughness on the flow above the spheres.

The regions of significant correlation between the flow field and particle force and torque fluctuations is of sizes comparable to the effective open channel height. For the correlation between particle quantities and streamwise velocity fluctuations, the regions of significant positive correlation are elongated in streamwise direction over several channel heights. The regions are flanked by regions of significant correlation with alternating signs at a spanwise distance comparable to the distance of high-speed streaks to low-speed streaks in smooth wall flows. The correlations of drag fluctuations and streamwise velocity fluctuations were compared to the correlations of spanwise torque and streamwise velocity fluctuations as well as to the correlations of drag fluctuations to pressure fluctuations. The results show that pressure is strongly correlated to drag fluctuations in both cases. The results can be interpreted in favour of the idea that pressure has a large effect on the characteristics of drag on a particle in the large sphere case and also a profound effect in the small sphere case, although here the hydraulically smooth regime is approached. The regions of significant correlations between flow field and spanwise force fluctuations or streamwise torque fluctuations are of sizes comparable to the channel height and exhibit complex shapes.

The overall picture is, that in the present cases force and torque on a fixed particle are related to velocity fluctuations in a region with a size comparable to the effective channel height. The time scales as well as the velocity scales related to force and torque are of order one when scaled with outer units. Future direct numerical simulations at higher Reynolds numbers would be beneficial to further address scaling aspects of the results above. Also experiments at high ratios of channel height to particle diameter would be needed to deepen the understanding of the correlation of turbulent flow to force and torque on particles.

Chapter 4

Onset of sediment erosion

Denn wir sind wie Baumstämme im Schnee. Scheinbar liegen sie glatt auf, und mit kleinem Anstoß sollte man sie wegschieben können. Nein, das kann man nicht, denn sie sind fest mit dem Boden verbunden. Aber sieh, sogar das ist nur scheinbar.

Die Bäume, Franz Kafka

In this chapter, the onset of sediment erosion is studied first by analysing the results of the flow configuration over fixed spheres, then by analysing direct numerical simulations with mobile eroding particles. Section 4.1 discusses possible criteria to predict sediment erosion by the fixed spheres results from chapter 3. Possible candidates are extreme events of drag, lift or spanwise torque. The outcome might depend on which criterion is chosen and, although studied in detail, it is not clear which quantity might be best suited to predict sediment erosion. Section 4.2 provides an analysis of the time-signals of particle force and torque conditioned to high drag, lift and spanwise torque and a discussion of the instantaneous and conditionally averaged flow fields related to high drag on a particle. Focus is given on the characterisation of events that might lead to sediment erosion. Simple probabilistic arguments are used in §4.3 to assess the potential of sediment erosion from the fixed sphere results and compare it to results from the literature. Direct numerical simulations of sediment erosion with mobile particles are presented in §4.4. The few simulations provide some insight into the mechanisms related to the onset of sediment erosion. The chapter closes with a summary, conclusion and recommendations for future work in §4.5.

4.1 Predicting sediment erosion from fixed sphere results

4.1.1 Failure modes of particle erosion

Simple predictions of the onset of sediment erosion by force and torque on a fixed particle might be based on exceeding a certain threshold of drag, lift or spanwise torque. More elaborate approaches consider combinations of force and torque on the particle related to geometrical considerations of the specific setup. Figure 4.1 shows a sketch of a particle resting centred on top of four particles in squared arrangement. This configuration corresponds to the configuration in the present case F10 and case F50. A failure scenario is that the particle on top of the spheres erodes along the plane perpendicular to the connection line between its centre (named C in figure 4.1) and the centres of the supporting particles (named B and F in figure 4.1). The failure mode is sketched by the dashed line in figure 4.1(a) with

(*a*) side view

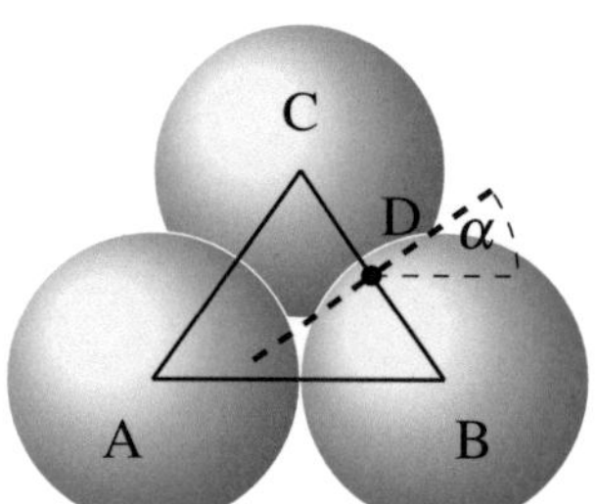

(*b*) top view

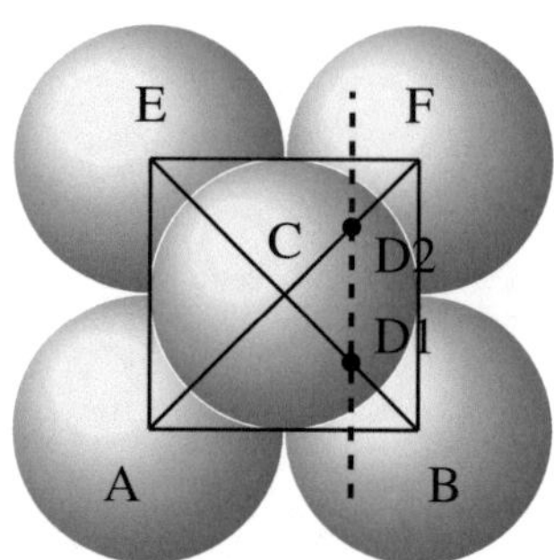

Figure 4.1: Sketch illustrating possible failure scenarios for a particle resting centred on top of four particles such that their particle centres A, B, C, E, F form a pyramid with a square base and equidistant side length. (*a*) Translational motion perpendicular to plane through particle centre C, B and F, here at an angle $\alpha = 35.3°$ to the horizontal plane (– – –). (*b*) Rotational motion around the axis through contact points D1 and D2 (– – –).

angle α to the horizontal axis. The onset of particle erosion might be predicted when the force in this direction, F_t, exceeds a certain threshold. This can be expressed by a combination of drag and lift via

$$F_t = F_x \cos(\alpha) + F_y \sin(\alpha) \geq F_{t,\text{thres}} , \tag{4.1}$$

where F_t is the force on a particle at angle α to the horizontal axis, here $\alpha = 35.3°$ for the given setup, and $F_{t,\text{thres}}$ is a given threshold. Another failure mode is that the particle rotates out of its resting position around the axis of support (dashed line through points of contact D1 and D2 in figure 4.1*b*). This leads to the definition of a torque, T_E, with respect to this axis by

$$T_E = T_z - F_x r_y^c - F_y r_x^c \leq T_{E,\text{thres}} , \tag{4.2}$$

where r_x^c and r_y^c are the streamwise and wall-normal distance of the contact points to the centre of the particle under consideration and $T_{E,\text{thres}}$ poses a threshold criterion. Other possible failure modes can be considered, e.g. rotation around support points in spanwise direction. However, the following discussion is limited to the analysis of F_x, F_y, T_z, F_t and T_E.

It seems reasonable to not only consider the force and torque on a particle, but also the impulse exerted on the particle. That is, in addition to the magnitude of a force or torque its duration might be of interest. Thus, another criterion to predict the onset of erosion based on fixed sphere results might use a certain threshold of impulse. The first study in line of such an approach could be considered to be Einstein & El-Samni (1949). The authors studied the mean duration of an approximated force above a certain threshold and found that it decreases exponentially as the threshold is increased. More recently, Celik *et al.* (2010) emphasised that the impulse of a flow event might be the important criteria to predict the onset of erosion. However, several difficulties are related with such a criteria. For example, the definition of an impulse from the time signal of a force requires the specification of either a threshold or an duration a priory. It is difficult to judge which choice might be best suited to predict the onset of sediment erosion as will discussed in more detail in section 4.1.3.

4.1.2 Equivalence of failure modes

In the previous section it was suggested to predict the onset of sediment erosion based on a threshold related to F_x, F_y, T_z, F_t and T_E. In the following the time signals of these quantities are compared to

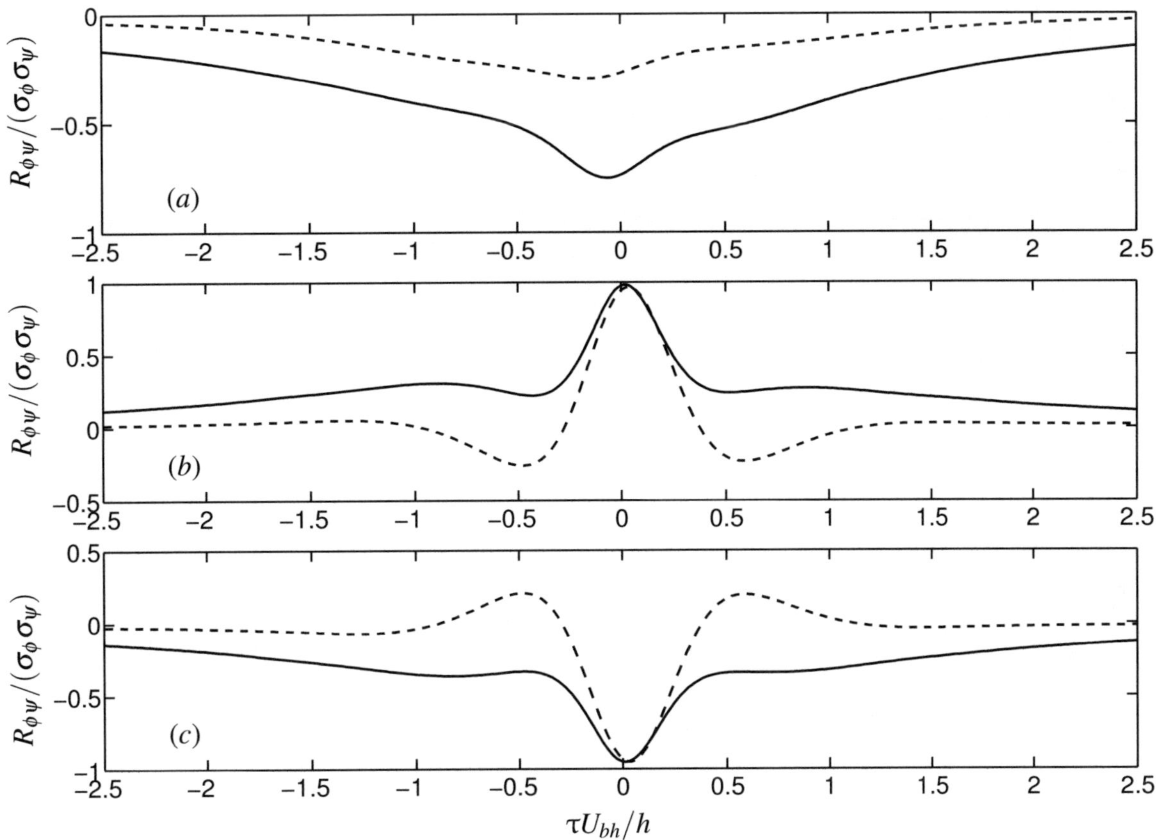

Figure 4.2: Cross-correlation as a function of time lag $\tau h/U_{bh}$ in case F10 (———) and F50 (– – –). (*a*) Drag fluctuations, F_x', and spanwise torque fluctuations, T_z', (*b*) drag fluctuations, F_x', and fluctuations of force along contact plane, F_t', (*c*) drag fluctuations, F_x', and fluctuations of torque around axis of contact points, T_E'.

assess which definitions would lead to equivalent predictions, which definition might lead to different results and which definition might not be relevant to define the onset of sediment erosion for the present cases. To this extent the cross-correlation in time, $R_{\phi\psi}(\tau)$, between the possible quantities are studied. The cross-correlations are defined by (3.14) with $\phi \neq \psi$. They are normalised in the following by the standard deviation of the respective quantities, i.e. σ_ϕ and σ_ψ.

The correlation function in time between drag and lift is discussed in section 3.5.5 (cf. figure 3.29). It is found that drag and lift correlate with a certain time lag $\tau U_{bh}/h$. The maximal correlation coefficient was found to be moderately high (0.45-0.55). Recall that here, U_{bh} is the bulk velocity based on the effective domain height, $h = H - y_0$, H is the domain height in wall-normal direction and $y_0 = 0.8D$ is the position of the virtual wall (cf. §3.1, §C.1). Figure 4.2(*a,b,c*) shows the correlation of drag fluctuations, F_x', with T_z', F_t' and T_E' respectively. Figure 4.2(*a*) reveals that the correlation of F_x' with T_z' decreases from case F10 to case F50 from a zero-lag coefficient of -0.74 to -0.27 respectively. In contrast to this, the correlations between F_x' and F_t' as well as the correlations between F_x' and T_E' (cf. figure 4.2*b,c*) exhibit magnitudes close to unity at $\tau = 0$ in both cases. This points at a close correlation between the signals and in particular at a strong influence of drag fluctuations on F_t' and T_E'.

Similar to the correlation of F_x' with T_z' (cf. figure 4.2*a*) the correlation of F_y' with T_z' shown in figure 4.3(*a*) decreases from case F10 to case F50. In contrast to figure 4.2(*a*) the maximum is found

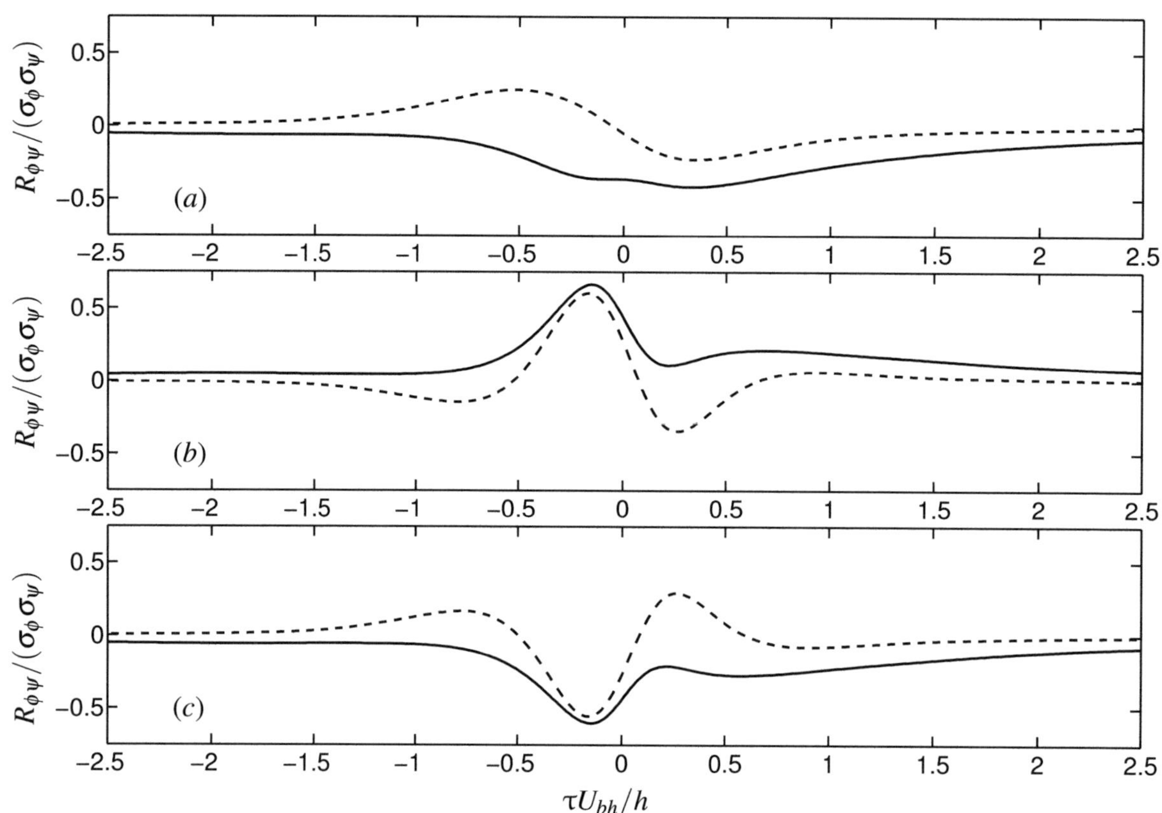

Figure 4.3: Cross-correlation as a function of time lag $\tau h/U_{bh}$ in case F10 (———) and F50 (– – –). (*a*) Lift fluctuations, F_y', and spanwise torque fluctuations, T_z', (*b*) lift fluctuations, F_y', and fluctuations of force along contact plane, F_t', (*c*) lift fluctuations, F_y', and fluctuations of torque around contact points, T_E'.

at non-zero values of τ. Thus while drag and spanwise torque fluctuations are in phase, lift and spanwise force fluctuations are not. This is similar to the correlation of drag and lift fluctuations shown figure 3.29. Even more similarities to the correlation of drag and lift – albeit with reverse axes – can be found for the correlation of F_y' with F_t' and F_y' with T_E' in figure 4.2(*b*,*c*). This suggests once more a strong influence of drag fluctuations on F_t' and T_E'.

Figure 4.4(*a*,*b*) shows the correlation of T_z' to F_t' and T_z' to T_E'. As F_t is a linear combination of drag and lift the strong decrease of the maxima in figure 4.4(*a*) from case F10 to F50 can be directly related to the strong decrease in the respective maxima in figure 4.2(*a*) and figure 4.3(*a*). T_E is a linear combination of drag, lift and spanwise torque and thus the strong decrease found in 4.4(*b*) as well as the similarities of the graphs to the correlation of drag fluctuations to spanwise torque in figure 4.2(*a*) might point at a weak contribution of spanwise torque fluctuations on T_E'. In general the drag fluctuations appear to be the dominant contributions to F_t' and T_E'. In particular, the contribution of lift fluctuations on F_t' and T_E' should lead to an asymmetric correlation function of F_x' and F_t' or T_E' in figure 4.2(*b*,*c*). However, the asymmetry of the correlation function in figure 4.2(*b*,*c*) is small which indicates a small contribution of lift to F_t' and T_E'.

It is shown in figure 3.29, that drag and lift fluctuations correlate. However, the correlation is complicated by a shift in time and as the magnitude of the correlation coefficient in case F10 and case F50 differs from unity, it is difficult to assess if a definition to predict sediment erosion by a threshold

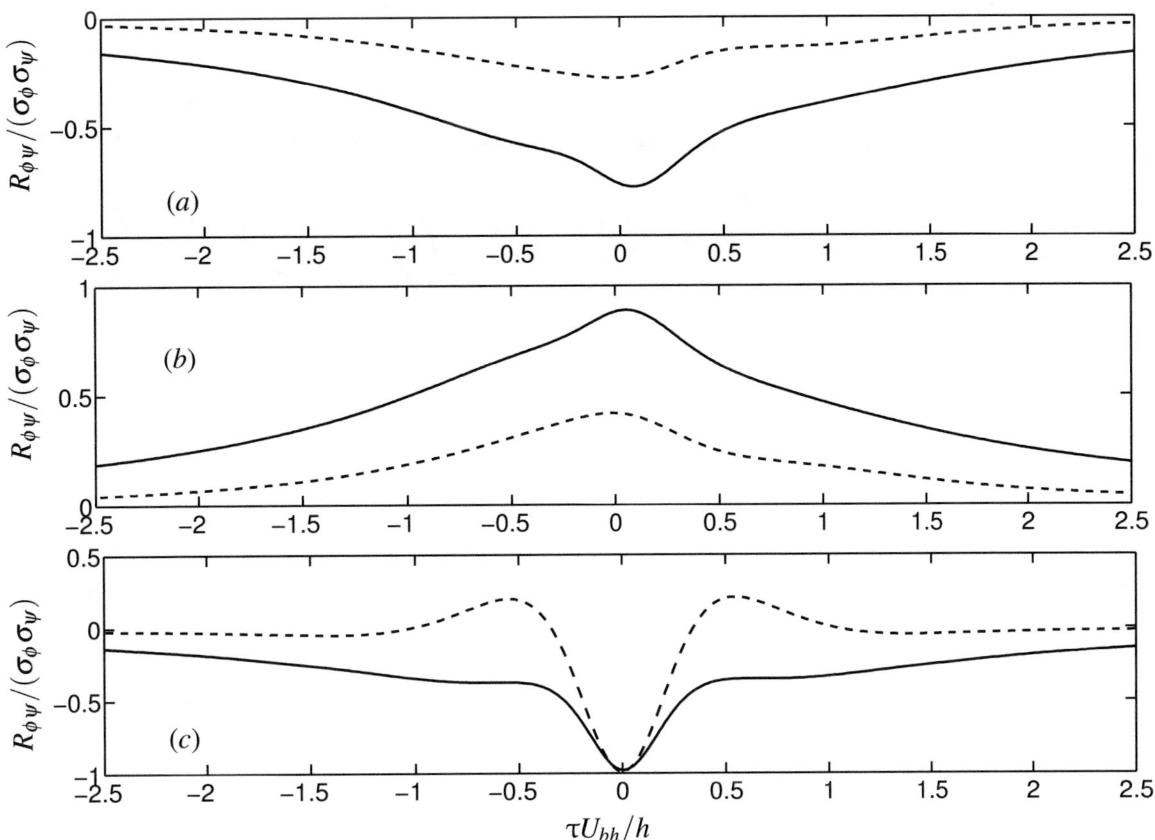

Figure 4.4: Cross-correlation as a function of time lag $\tau h/U_{bh}$ in case F10 (———) and F50 (– – –). (*a*) Spanwise torque fluctuations, T_z', and fluctuations of force along contact plane, F_t', (*b*) spanwise torque fluctuations, T_z', and fluctuations of torque around contact points, T_E', (*c*) fluctuations of torque around contact points, T_E', and fluctuations of force along contact plane, F_t'.

of high drag leads to a similar result as a criterion based on high lift. To further address this question, figure 4.5 provides the joint probability density functions of drag and lift fluctuations at a shift in time that maximises the correlation shown in figure 3.29. The overall picture is that low (high) drag events correlate with low (high) lift events, respectively. However, for a given level of drag fluctuations a rather broad range of lift fluctuations is present and vice versa. Thus it is likely that a definition to predict the onset of sediment erosion based on high drag will lead to somewhat different results than a definition based on high lift.

From the above analysis several conclusions can be drawn with respect to predicting the onset of sediment erosion by force and torque on particles. A definition based on F_t or T_E most likely leads to equivalent results as definition based on high drag. The correlation coefficient between drag and spanwise torque is non-zero but decreases from case F10 to case F50. Thus an equivalence between the predictions based on drag or spanwise torque is difficult to establish. From the dominance of the drag fluctuations on T_E' one might speculate that spanwise torque is of minor relevance to define the onset of erosion. A correlation between drag and lift fluctuations exists, albeit at a lag in time. The joint probability density functions reveal that a wide range of drag fluctuations exist for a given

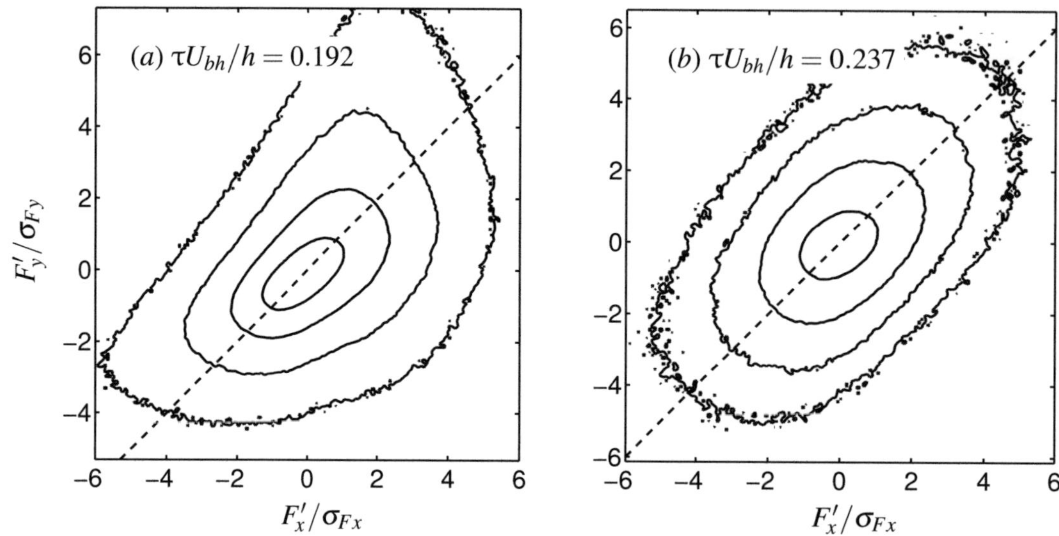

Figure 4.5: Joint probability density function of drag and lift fluctuations with a time-lag that maximises the drag-lift cross-correlation in time. Lines show iso-contours at a level of probability equal to $[1e-4\ 1e-3\ 1e-2\ 1e-1]$. (a) Case F10 at $\tau U_{bh}/h = 0.192$, (b) case F50 at $\tau U_{bh}/h = 0.237$. Dashed lines are added to guide the eye and show relation $F'_y/\sigma_{Fy} = F'_x/\sigma_{Fx}$.

threshold of lift fluctuations and vise versa. Thus a criterion to define the onset of erosion based on high drag is not necessarily equivalent to a criterion based on high lift.

4.1.3 Choice of threshold value

The focus of the present section is to clarify if in case F10 and case F50 there is a threshold value of drag, lift or spanwise torque that might be particularly suited to predict the onset of sediment erosion. Here, the answer is approached by looking at statistical properties of time intervals of signals conditioned to a threshold criteria.

The time intervals are defined as

$$I_t(\phi', \phi_{\text{thres}}) = (t_1, t_2) \qquad \text{with } \phi'(t) \geq \phi_{\text{thres}} \forall t \in I_t \text{ and } \phi'(t_1), \phi'(t_2) < \phi_{\text{thres}}\,, \tag{4.3}$$

were ϕ is the chosen defining particle quantity, i.e. drag ($\phi' = F'_x$), lift ($\phi' = F'_y$) or negative spanwise torque ($\phi' = -T'_z$), t_1 and t_2 define the interval, and ϕ_{thres} is the respective threshold criteria. The duration of an interval I_t is defined as the integral $\tau_I = \int_{I_t} \mathrm{d}t = t_2 - t_1$. The integral of the quantity ϕ over the interval I_t is defined as

$$\mathscr{I}_\phi(\phi, \phi_{\text{thres}}) = \int_{I_t} \phi \,\mathrm{d}t\,. \tag{4.4}$$

The integral $\mathscr{I}$ is called impulse in the following.

Before analysing the time signals of drag, lift and spanwise torque on the particle in case F10 and case F50 it is helpful to consider the characteristics of the definition above in more detail. Figure 4.6 illustrates the definitions by a short sequence of the drag fluctuation on a single particle in case F10 around the local maximum at $tU_{bh}/h = 213.5$. Two time intervals, I_t^0 and I_t^2, are shown, defined by a threshold of $\phi_{\text{thres}}/\sigma_\phi = 0$ and 2 respectively. Note, that the time signal ϕ' in the interval I_t^2 exhibits

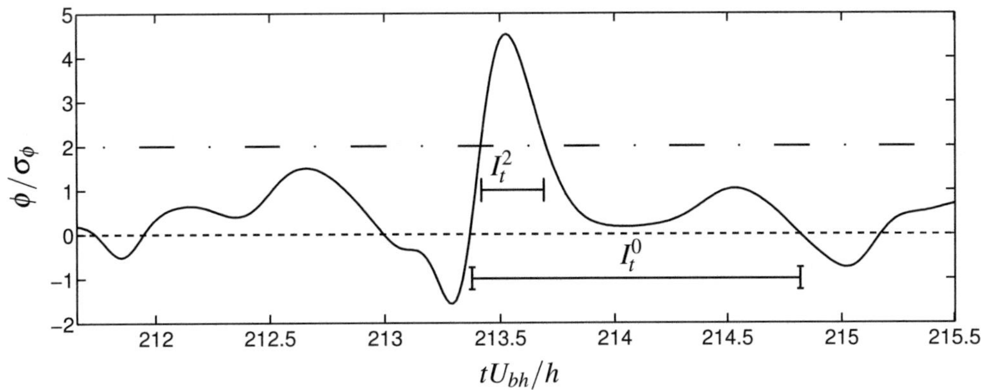

Figure 4.6: Sketch of definition of intervals, figure shows time signal of ϕ (———) as a function of time jointly with two thresholds $\phi_{\text{thres}}/\sigma_\phi = 0$ (– – –) and $\phi_{\text{thres}}/\sigma_\phi = 2$ (– · –) defining two intervals including the maximum at $tU_{bh}/h = 213.5$.

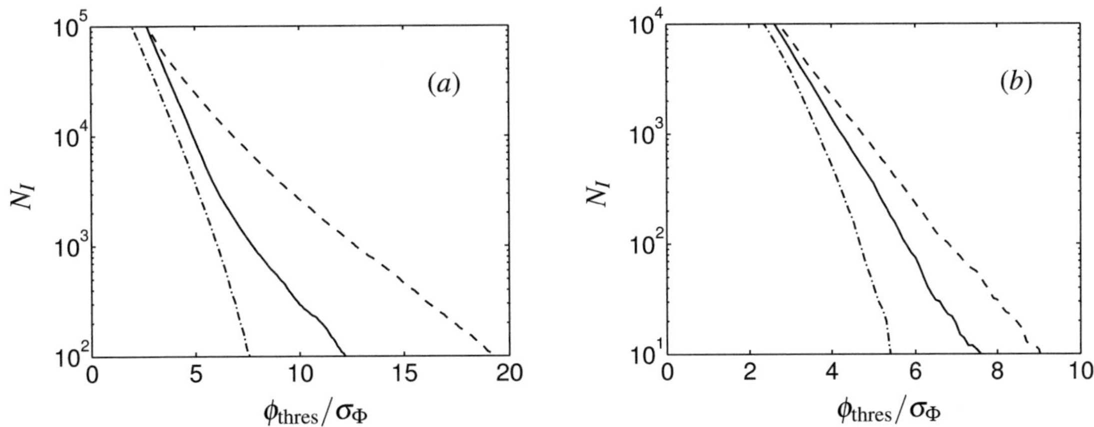

Figure 4.7: Number of time intervals I_t, N_I, as defined in (4.3) in case F10 (*a*) and case F50 (*b*). The lines show results when the criteria for ϕ' equals to drag fluctuations (———), lift fluctuations (– – –) and spanwise torque fluctuations (– · –).

a single pronounced maximum whereas the time signal ϕ' exhibits two local maxima. As the value of ϕ_{thres} is gradually increased from a threshold value of $\phi_{\text{thres}}/\sigma_\phi = 0$, the interval I_t^0 will eventually be replaced by two time intervals of a much smaller (about half) duration τ_I. Further increasing the value of $\phi_{\text{thres}}/\sigma_\phi$ reduced the duration of the time interval τ_I only gradually. Eventually the number of intervals is reduced, until only intervals with a pronounced isolated maximum, such as seen for I_t^2, remain. In the extreme case of $\phi_{\text{thres}} < \min(\phi)$ only one interval I_t is identified which has the duration of the observation time τ_c of the signal ϕ and several local maxima. In the extreme case of $\phi_{\text{thres}} > \max(\phi)$ no interval is defined.

Figure 4.7 shows the number of intervals, N_I, obtained from the time signals of drag, lift and spanwise torque in case F10 and case F50 for $\phi_{\text{thres}}/\sigma_\phi > 0$. In the given range, the number of intervals N_I decrease approximately exponentially. For some quantities a change in slope can be recognised, e.g. for drag fluctuations in case F10 at about $\phi_{\text{thres}}/\sigma_\phi \approx 7$. The number of intervals in

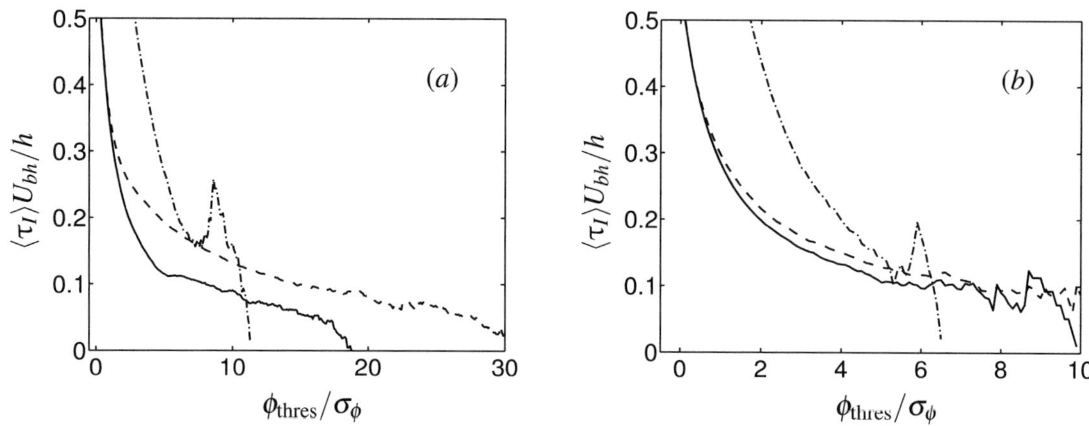

Figure 4.8: Average duration of time intervals I_t, as a function of ϕ_{thres}. Line style and panels as in figure 4.7.

case F10 is found to be higher than in case F50. This is due to two reasons, on one hand in case F10 about 9 times more particles are considered, on the other hand, the time signals in case F10 are more intermittent (cf. figure 3.17 and figure 3.20) and thus the number of intervals for extreme events with $\phi_{\text{thres}}/\sigma_\phi > 5$ is larger in case F10.

Figure 4.8 shows the mean duration of the interval obtained from the signals of case F10 and case F50. The mean duration of drag or lift intervals, $\langle\tau_I\rangle$, drops rapidly for small values of $\phi_{\text{thres}}/\sigma_\phi$. For $\phi_{\text{thres}}/\sigma_\phi \geq 5$ ($\phi_{\text{thres}}/\sigma_\phi \geq 2$) the drop in case F10 (F50) is more gradual and exhibits a value in the range of 0.1 to 0.2 h/U_{bh}. The mean duration of spanwise torque intervals does not exhibit a similar flattening of the profile. Moreover, in that case the accuracy of the measure is considerably reduced above $\phi_{\text{thres}}/\sigma_\phi > 5.5$ due to lack of samples (cf. figure 4.7).

Following the discussion related to figure 4.6 the range of a more gradual decrease of $\langle\tau_I\rangle U_{bh}/h$ in figure 4.8 can be interpreted as the range of $\phi_{\text{thres}}/\sigma_\phi$ that defines intervals with mostly one single, well pronounced maximum. Some support for this interpretation comes from the value of the mean duration, $\langle\tau_I\rangle$, which is in the range of 0.1 to 0.2 h/U_{bh} and thus smaller than the micro-scale and smaller than the position of the local minimum observed in the auto-correlation of drag and lift on the particles (cf. table 3.9). Thus, it seems reasonable to assume that the intervals indeed contain mostly only a single isolated maximum.

The above suggests that using a threshold of $\phi_{\text{thres}}/\sigma_\phi > 5$ (2) in case F10 (F50) for drag and lift might be sufficient to identify extreme events related to a pronounced isolated maximum within the time signal of an average duration of 0.1 to 0.2 h/U_{bh}. In case of spanwise torque the picture is less clear. From figure 3.20(*a*,*b*) and figure 4.7 it can be concluded that a threshold of $\phi_{\text{thres}}/\sigma_\phi > 5$ is related to rare events. Figure 4.8 shows that in that case the related mean durations are similar to the mean duration of drag and lift. A similar flattening of the mean duration profiles however does not occur. This could be seen as an indication, that in case of spanwise torque the intervals do on average not contain a similarly pronounced maximum as for the forces.

Figure 4.9 shows the mean value of the impulse on a particle as defined by (4.4) to investigate if a threshold exists for which the impulse is on average at a maximum. The impulse is normalised by $\mathscr{I}_R = F_R h/U_{bh}$ in case of drag and lift fluctuations and by $\mathscr{I}_R = T_R h/U_{bh}$ in case of negative spanwise torque fluctuations. The profile related to spanwise torque fluctuations in figure 4.9 exhibits local maxima. One of the maxima is related to a threshold criteria close to zero in both flow cases and thus

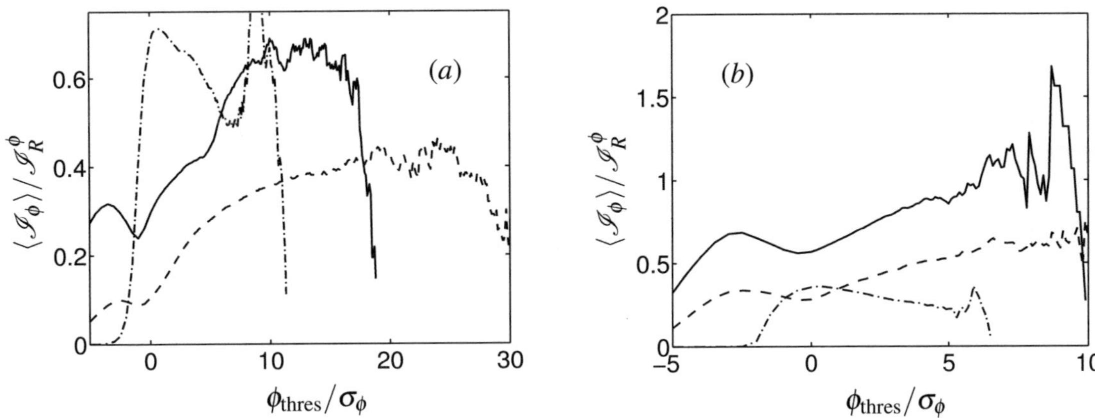

Figure 4.9: Mean value of $\mathscr{I}_\phi$ defined by (4.4) as a function of threshold $\phi_{\text{thres}}/\sigma_\phi$ normalised by $\mathscr{I}_R = F_R h/U_{bh}$ in case of force and by $\mathscr{I}_R = T_R h/U_{bh}$ in case of torque. Line style and panels as in figure 4.7.

does not provide a definition for a conclusive threshold criteria. A second local maximum exists at values around $\phi_{\text{thres}}/\sigma_\phi \approx 9$ (6) in case F10 (F50). However, the number of samples are small (cf. figure 4.7) and thus the maximum cannot be expected to be related to a conclusive threshold criteria. For drag or lift, a local maximum is visible at negative values of ϕ_{thres} and a local minimum occurs at $\phi_{\text{thres}}/\sigma_\phi \approx 0$. For $\phi_{\text{thres}} > 0$ the values of $\langle \mathscr{I}_\phi \rangle / \mathscr{I}_R^\phi$ first increase. The increase of $\mathscr{I}/\mathscr{I}_R$ with increasing ϕ_{thres}/ϕ indicates that the mean value of ϕ over the time defined by the interval increases on average more rapidly than the mean duration of the intervals $\langle \tau_I \rangle$ decreases. Maxima exist around $\phi_{\text{thres}}/\sigma_\phi \approx 10$ (8) for drag and around $\phi_{\text{thres}}/\sigma_\phi \approx 25$ (9) for lift in case F10 (F50). As in the case of spanwise torque, it is difficult to assess if these maxima can serve as a physically motivated definition of a threshold criteria as they might represent a bias due to a lack of samples (cf. figure 4.7). The above discussion of figure 4.9 shows the limitations of impulse as defined by (4.4) to predict the onset of sediment transport for the present cases.

4.2 Conditioned events in fixed sphere case

4.2.1 Conditionally averaged time signals

This section presents conditionally averaged time signals of force and torque fluctuations on fixed particles. The averaging conditions are events with drag, lift and spanwise torque fluctuations larger than five standard deviations. The conditional averaging operator is defined as

$$\langle \psi \rangle_{\text{cond}}(\tau) = \frac{1}{N_{\text{cond}}} \sum_{i=1}^{N_t} \sum_{l=1}^{N_p} \psi^l(t_i + \tau) \quad \forall i, l \text{ for which } \phi^l(t_i) > \phi_{\text{thres}}\,, \tag{4.5}$$

were $\langle \psi \rangle_{\text{cond}}$ is the conditionally averaged particle quantity as a function of time lag, τ, ψ^l and ϕ^l are particle quantities of particle l, where ϕ is the conditioning quantity with a conditioning threshold ϕ_{thres}, N_t is the number of time steps, N_p is the number of particles under consideration and N_{cond} is the number of realisations that meet the condition $\phi^l(t_i) > \phi_{\text{thres}}$.

Figure 4.10 shows the conditionally averaged drag, lift and spanwise torque fluctuations normalised by the standard deviation of drag, lift and spanwise torque respectively. The averaging condition is drag with $\phi = F_x' > 5\sigma_{Fx} = \phi_{\text{thres}}$. This criterion is large enough to obtain mostly single events

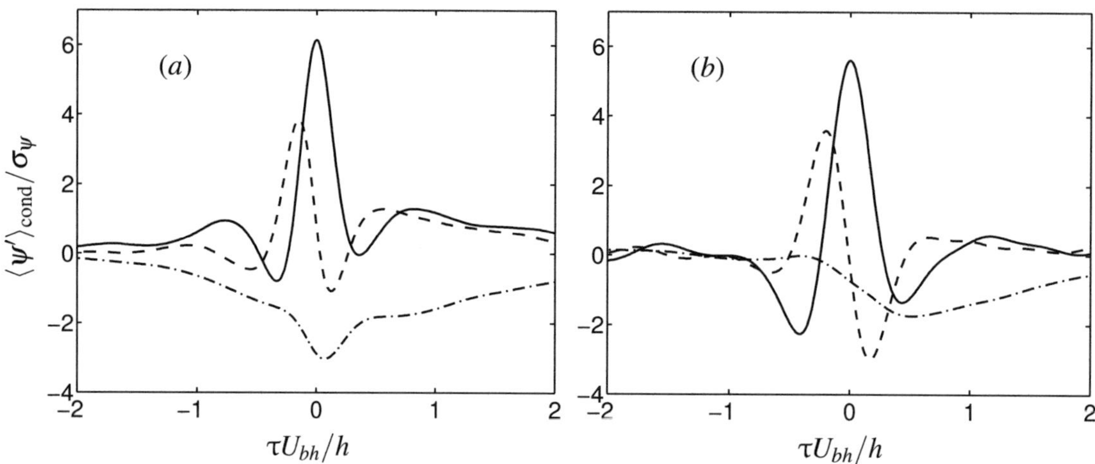

Figure 4.10: Conditionally averaged drag (———), lift (– – –) and spanwise torque (– · –) fluctuation in case F10 (*a*) and case F50 (*b*) normalised by the respective standard deviation. Averaging condition is drag with a threshold criterion of $F_x' > 5\sigma_{Fx}$.

of high drag fluctuations (cf. discussion in §4.1) and small enough to obtain enough samples to reach converged statistics (cf. figure 4.7).

The conditionally averaged profile of drag fluctuations, $\langle F_x'\rangle_{\mathrm{cond}}/\sigma_{Fx}$, in figure 4.10 exhibits a maximum of $\langle F_x'\rangle_{\mathrm{cond}}/\sigma_{Fx} = 6.2$ (5.6) in case F10 (F50) at $\tau = 0$. From this maximum the profile drops rapidly towards local minima up to values of $\langle F_x'\rangle_{\mathrm{cond}}/\sigma_{Fx} = -0.8$ (-2.3) in case F10 (F50). The duration of the high drag event from minimum to minimum is about $\tau U_{bh}/h = 0.71$ (0.85) in case F10 (F50) which is a bit less than twice the time-lag from maximum to minima in the auto-correlations, $\tau_{\min}$ (cf. table 3.9). Similar to the drag fluctuations the conditionally averaged profiles of lift fluctuations in figure 4.10, $\langle F_y'\rangle_{\mathrm{cond}}/\sigma_{Fy}$, exhibit pronounced maxima, albeit at a shift in time, i.e. the maximum in case F10 (F50) is located at $\tau U_{bh}/h = -0.15$ (-0.20) with a value of $\langle F_y'\rangle_{\mathrm{cond}}/\sigma_{Fy} =$ 3.9 (3.6). Although, this is lower than the maxima of $\langle F_x'\rangle_{\mathrm{cond}}/\sigma_{Fx}$, the related lift events can be classified as very high with less than 1% probability of occurrence (cf. figure 3.17). The minima of $\langle T_z'\rangle_{\mathrm{cond}}/\sigma_{Tz}$ in figure 4.10 are weaker and exceed values of $\langle T_z'\rangle_{\mathrm{cond}}/\sigma_{Tz} = -3.0$ at $\tau U_{bh}/h =$ 0.1 in case F10 and $\langle T_z'\rangle_{\mathrm{cond}}/\sigma_{Tz} = -1.7$ at $\tau U_{bh}/h = 0.5$ in case F50. This shows once more that the correlation between drag and spanwise torque fluctuations decreases from case F10 to case F50, similar to the cross-correlations in figure 4.2(*a*).

It is interesting to note, that the shape of the conditionally averaged spanwise torque fluctuations differ in its characteristics from case F10 to F50. While in case F10 the curve is approximately symmetric with respect to the minimum, in case F50 the profile is strongly asymmetric with respect to the minimum. As a consequence in case F10 the location of extreme spanwise torque fluctuation coincides with the location of high drag fluctuations. While in case F50, the location of extreme spanwise torque fluctuations coincides with the location of a local minima of drag fluctuations.

It is remarkable, that the shape of the conditionally averaged quantities show similar features as the respective auto-correlation and cross-correlations between the conditioning quantity and the conditionally averaged quantity (i.e. figure 3.26*a*, figure 3.29 and figure 4.2*a*) albeit with a reversed τ-axis. A possible explanation of the agreement might be a high importance of the extreme events in the respective correlation functions. Another explanation might be that the shape of force and torque events is similar independently of the magnitude of the force.

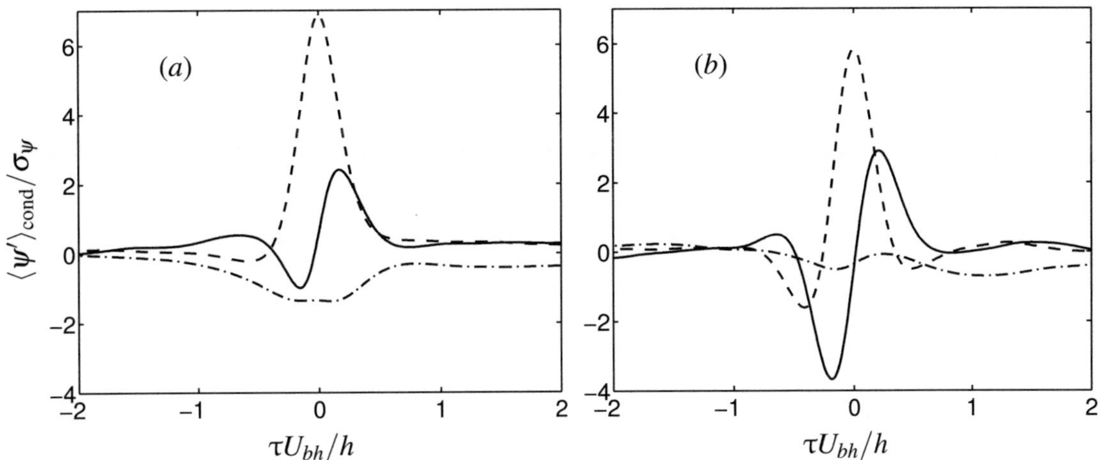

Figure 4.11: Conditionally averaged drag (———), lift (– – –) and spanwise torque (– · –) fluctuation in case F10 (*a*) and case F50 (*b*) normalised by the respective standard deviation. Averaging condition is lift with a threshold criterion of $F_y' > 5\sigma_{Fy}$.

Figure 4.11 displays the conditionally averaged drag, lift and spanwise torque fluctuations conditioned to lift with a threshold criterion of $F_y' > 5\sigma_{Fy}$. This criterion guarantees sufficient samples as well as high lift events with pronounced maxima (cf. §4.1). However, note that the criterion might be seen as less critical than that in figure 4.10 as lift is more intermittent (cf. figure 3.17). The maximum of $\langle F_x'\rangle_{\rm cond}/\sigma_{Fx}$ has a value of 2.4 (2.9) in case F10 (F50) at a time-lag $\tau U_{bh}/h = 0.2$. Note that these values are lower than the respective maxima of lift fluctuation in figure 4.10. The maximum of $\langle F_y'\rangle_{\rm cond}/\sigma_{Fy}$ is at a value of 6.9 (5.8) in case F10 (F50) and thus compares to the maxima of the conditioning quantity in figure 4.10. The findings for drag and lift as a conditioning criterion suggest, that the profile of the quantity used for the conditioning exhibits commonly the highest values. The profiles of the other conditionally averaged quantities exhibit lower values. This result can be linked to the broad shape of the two-dimensional probability density function (e.g. figure 4.5). In contrast to figure 4.10 the profiles of $\langle T_z'\rangle_{\rm cond}/\sigma_{Tz}$ in figure 4.11 do not show pronounced minima and are only weakly correlated in case F50. Once more the shape of the conditionally averaged profiles discussed above show similar features as the respective auto- and cross-correlations of conditioning and averaged particle quantity in figure 3.26(*b*), figure 3.29 and figure 4.3(*a*).

Figure 4.12 shows the conditionally averaged profiles of drag, lift and spanwise torque fluctuation conditioned to spanwise torque with a threshold criterion of $T_z' > 5\sigma_{Tz}$. In case F10 the high spanwise torque events relate to high events of drag and lift forces reaching values of $\langle F_x'\rangle_{\rm cond}/\sigma_{Fx} = 4.0$ and $\langle F_y'\rangle_{\rm cond}/\sigma_{Fy} = 2.6$. The conditionally averaged particle statistics compare to the cross-correlation (figure 4.2*a*, 4.3*a*, reverted τ-axis) and the auto-correlation (figure 3.24*c*). In case F50 the picture is less clear, due to less satisfactory statistical convergence. However, the maxima appear to be at lower values. This is confirmed by comparing the results at a somewhat lower threshold criterion of $T_z' > 4\sigma_{Tz}$ for which the statistical convergence is much better (plot omitted).

For completeness, figure 4.13 compares the profiles of $\langle F_x'\rangle_{\rm cond}$, $\langle F_t'\rangle_{\rm cond}$ and $\langle T_E'\rangle_{\rm cond}$ conditioned to drag with a threshold $F_x' > 5\sigma_{Fx}$. The time signals are very similar when conditioned to high drag and support that, for the current cases, a prediction of sediment erosion based on F_t and T_E leads to equivalent results than a definition based on drag fluctuations.

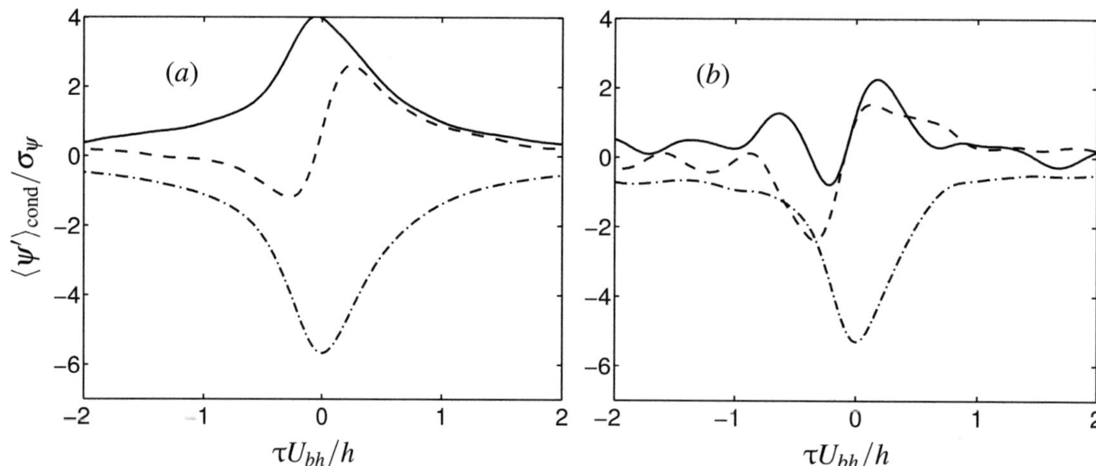

Figure 4.12: Conditionally averaged drag (———), lift (– – –) and spanwise torque (– · –) fluctuation in case F10 (*a*) and case F50 (*b*) normalised by the respective standard deviation. Averaging condition is spanwise torque with a threshold criterion of $T_z' \leq -5\sigma_{Tz}$.

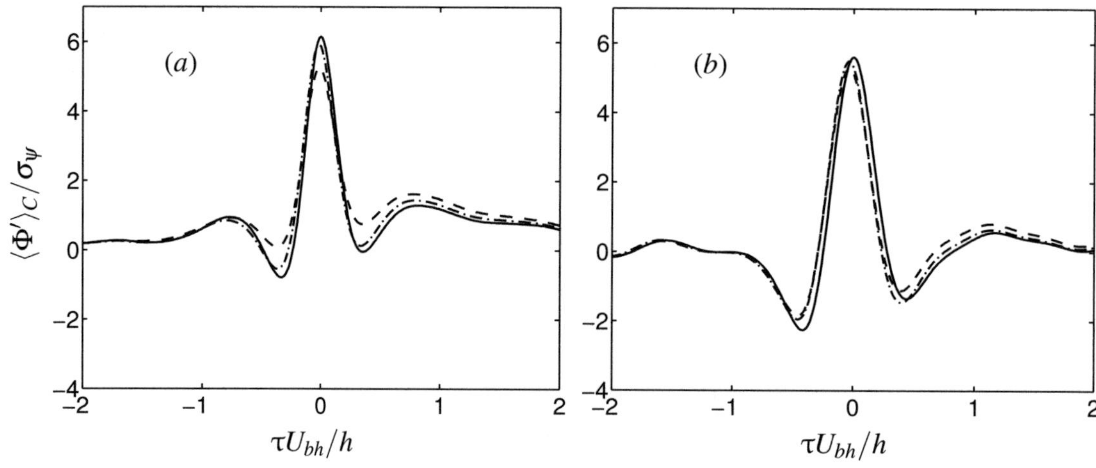

Figure 4.13: Conditionally averaged fluctuations of drag (———), tangential force (– – –) and moment around contact point (– · –) in case F10 (*a*) and case F50 (*b*) normalised by the respective standard deviation. Averaging condition is drag with a threshold criterion of $F_x' > 5\sigma_{Fx}$. Note that the sign of the fluctuations of the moment around the contact point was changed.

4.2.2 Conditioned instantaneous flow structures

Figures 4.14 to 4.17 show instantaneous flow structures centred to particles that exceed the threshold of $F_x' > 5\sigma_{Fx}$. The criterion to visualise the flow structures in the figures is chosen high, such that the structures in the figures are small and well defined. The conditioned flow fields of case F10 shown in figure 4.14 differ from those of case F50 shown in figure 4.16 in several respects. One difference is the size of the spheres compared to the size of the flow structures. In case F10 the spheres are small with respect to the flow structures while in case F50 the sizes are comparable. This might explain that in case F50 the flow structures appear to be somewhat more affected by the presence of the spheres than in case F10. A related difference is the correlation of particle drag fluctuations over a range of

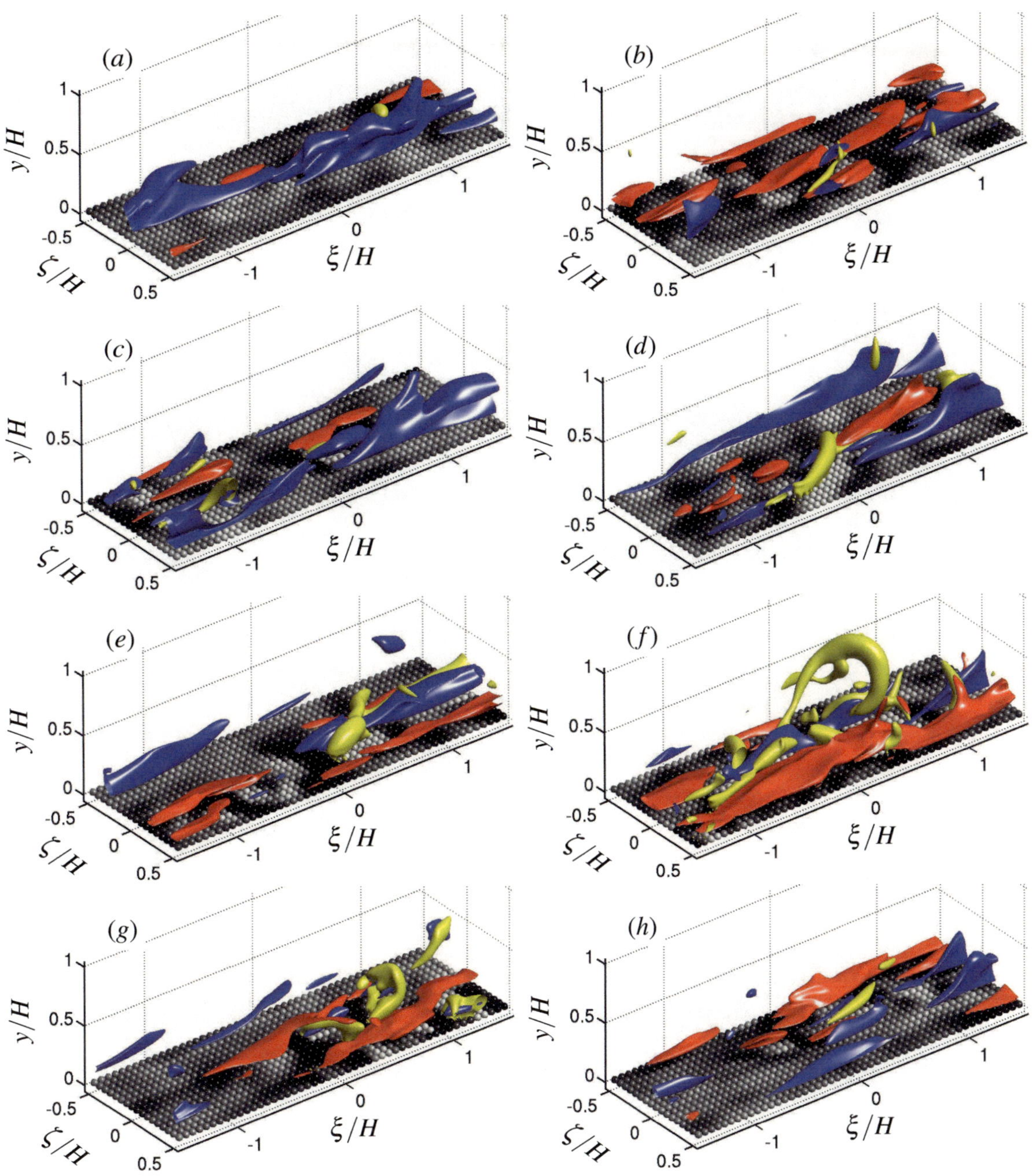

Figure 4.14: Instantaneous flow field in case F10 centred around particles with $F_x' > 5\sigma_{Fx}$. Surfaces show iso-surfaces of u' equal to $4u_\tau$ ($-4u_\tau$) in red (blue) as well as iso-surfaces of pressure fluctuations of $p'/(\rho_f u_\tau^2) = -7$ in yellow. Particles are coloured from white to black according to the respective value of F_x'/σ_{Fx} in the range of -2 to 2.

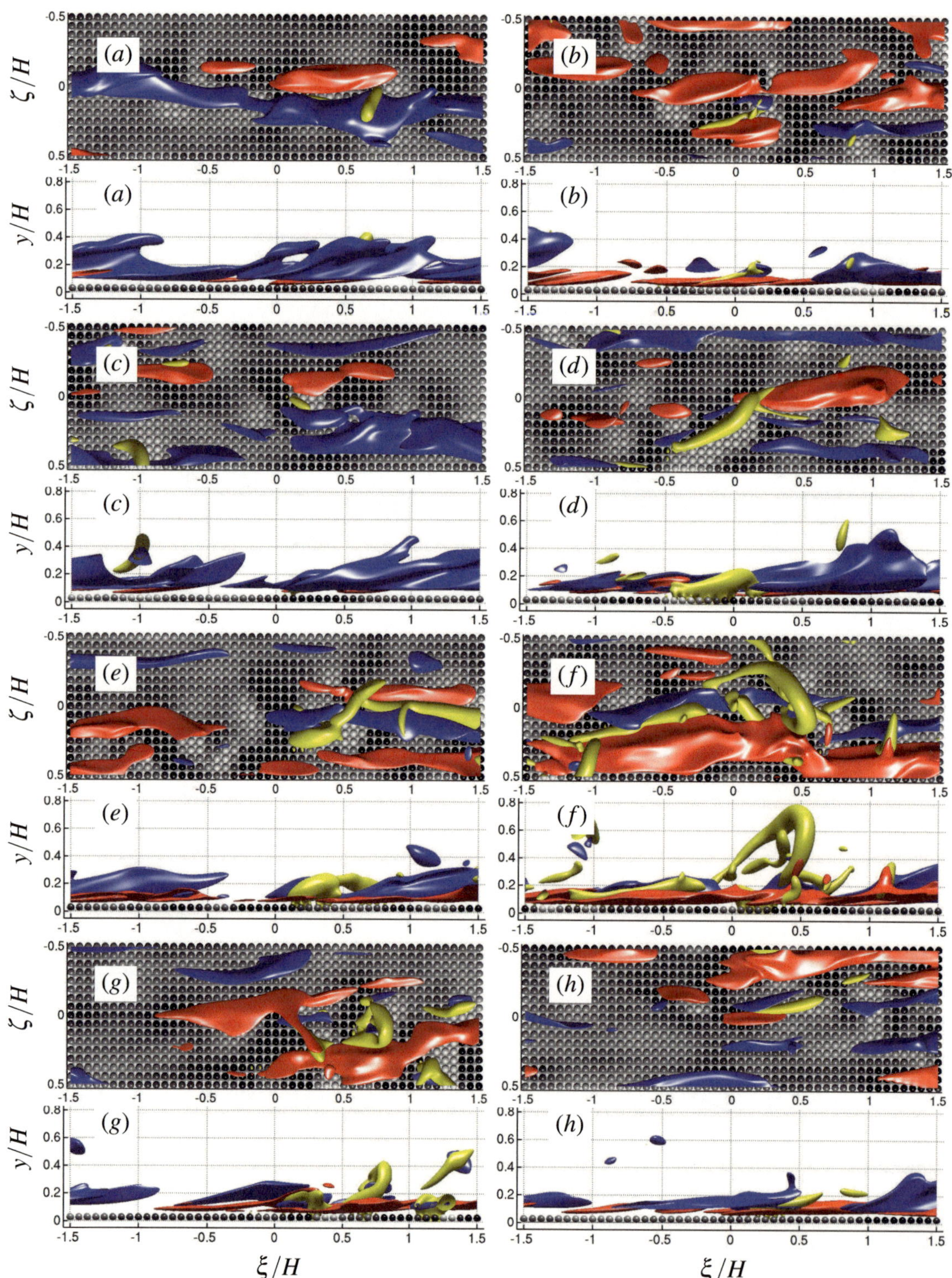

Figure 4.15: Instantaneous flow fields as shown in figure 4.14 in top and side view.

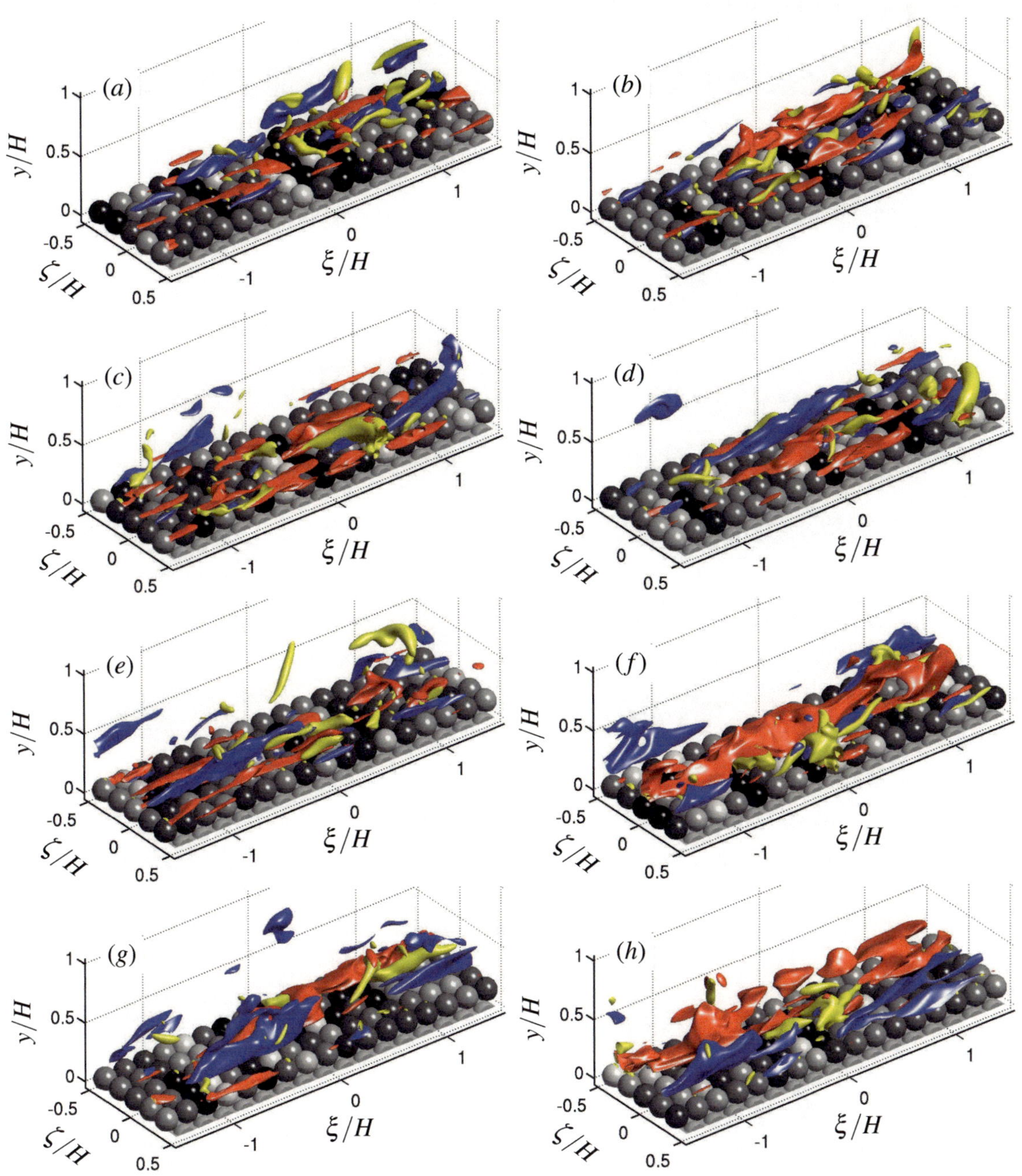

Figure 4.16: As in figure 4.14 but for case F50.

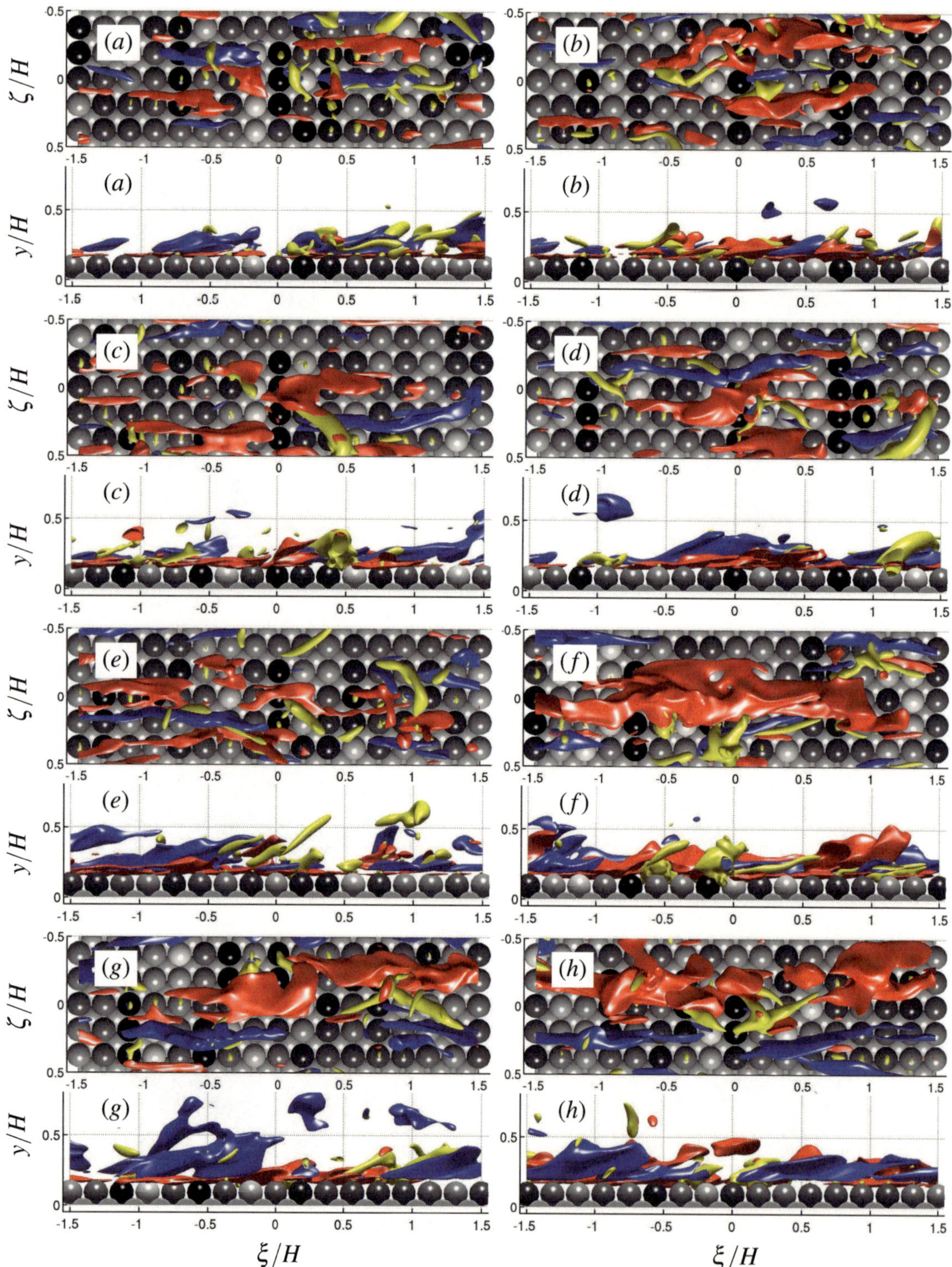

Figure 4.17: Instantaneous flow fields as shown in figure 4.16 in top and side view.

particle. In both figures, the spheres are coloured according to their instantaneous drag fluctuations. In case F10 drag fluctuations of the particles are well correlated in space over several particles which is not visible to the same extent in case F50. A straightforward relation between the flow structures in the figures and particle drag does not seem to exist. However, some observations can be made.

In case F10, figures 4.14(*a,b,c,d,h*) and the respective panels in figure 4.15 show high instantaneous streamwise velocity fluctuations in the vicinity of the particle with $F_x' > 5\sigma_{Fx}$ positioned at ($\xi = 0$, $y = D/2$, $\zeta = 0$). This is in agreement with the idea that positive streamwise velocity fluctuations in the vicinity of the particle will relate to positive drag when their streamwise length is equal or larger than the particle diameter. Contrarily, figures 4.14(e, f, g) and the respective panels in figure 4.15 reveal negative velocity fluctuation in the direct vicinity of the particle. High negative pressure fluctuations are present near-by the particle in all panels in figures 4.14 and 4.15. These fluctuations have shapes that agree with the notation of hairpin vortices as described for example by Robinson (1991). The region in the vicinity of the particle is highly active in most cases and might be linked to the process of a so called burst event (cf. Rao *et al.*, 1971; Jiménez *et al.*, 2005; Sheng *et al.*, 2009; Jiménez & Kawahara, 2012). This aspect is particularly interesting since burst events are thought to be related with sediment erosion since the paper by Jackson (1976). A more recent discussion on the relation of burst events to high shear events in a smooth wall channel flow at $\mathrm{Re}_\tau = 1470$ can be found in Sheng *et al.* (2009).

Despite some differences the over all picture in case F50 is similar to those in case F10. Figures 4.16 and 4.17 show high negative pressure fluctuations in the vicinity of the particle exceeding the threshold criterion analogously to case F10. In particular on the downstream side of the particle with $F_x' > 5\sigma_{Fx}$ high negative pressure fluctuation seem to occur. Also, the particles are located in highly turbulent regions in all instantaneous flow fields shown. With a few exceptions the particle is found next to high streamwise velocity fluctuation.

4.2.3 Conditionally averaged flow structures

The discussion above shows that the instantaneous flow field around the particles subject to high drag is complex and highly turbulent. A straightforward relation of the instantaneous flow structures to events of high drag does not seem to exist. Therefore, in this section focus is given to study the relation of flow structures to high drag by conditionally averaged flow fields. The averaging operator is defined analogously to (4.5) as

$$\langle \psi \rangle_{\mathrm{cond}}(\xi, y, \zeta) = \frac{1}{N_{\mathrm{cond}}} \sum_{i=1}^{N_t} \sum_{l=1}^{N_p} \psi^l(x_p + \xi, y, z_p + \zeta, t_i) \quad \forall i, l \text{ for which } \phi^l(\mathbf{x}_p, t_i) > \phi_{\mathrm{thres}}\,, \qquad (4.6)$$

where $\langle \psi \rangle_{\mathrm{cond}}(\xi, y, \zeta)$, is the conditionally averaged flow field, $\psi(\mathbf{x}, t)$ is a fluid field fluctuation quantity, $\phi^l(\mathbf{x}_p, t)$ is the conditioning quantity of particle l located at $\mathbf{x}_p = (x_p, y_p, z_p)$, ξ and ζ are the spatial shifts with respect to the particle centre, ϕ_{thres} is the threshold of the conditioning criterion, N_t is the number of time steps t_i, N_p is the number of particles l under consideration and N_{cond} is the number of realisations that meet the condition $\phi^l(t_i) > \phi_{\mathrm{thres}}$. Averaging was carried out over the saved instantaneous flow fields and particle data in case F10 and case F50. The averaging condition was drag with a threshold $F_x' > 5\sigma_{Fx}$. This can be regarded as a very high (cf. discussion in §4.1). Some of the instantaneous flow fields related to such a threshold are visualised in figures 4.14 to 4.17 above. The total number of samples is $N_{\mathrm{cond}} = 545$ (32) in case F10 (F50) from 56 (24) snapshots within an interval of 52 U_{bH}/H (118 U_{bH}/H). Note, that in case F10 the number of samples is higher due to the higher number of particles in one snapshot (cf. table 3.1). Additionally, in this case the

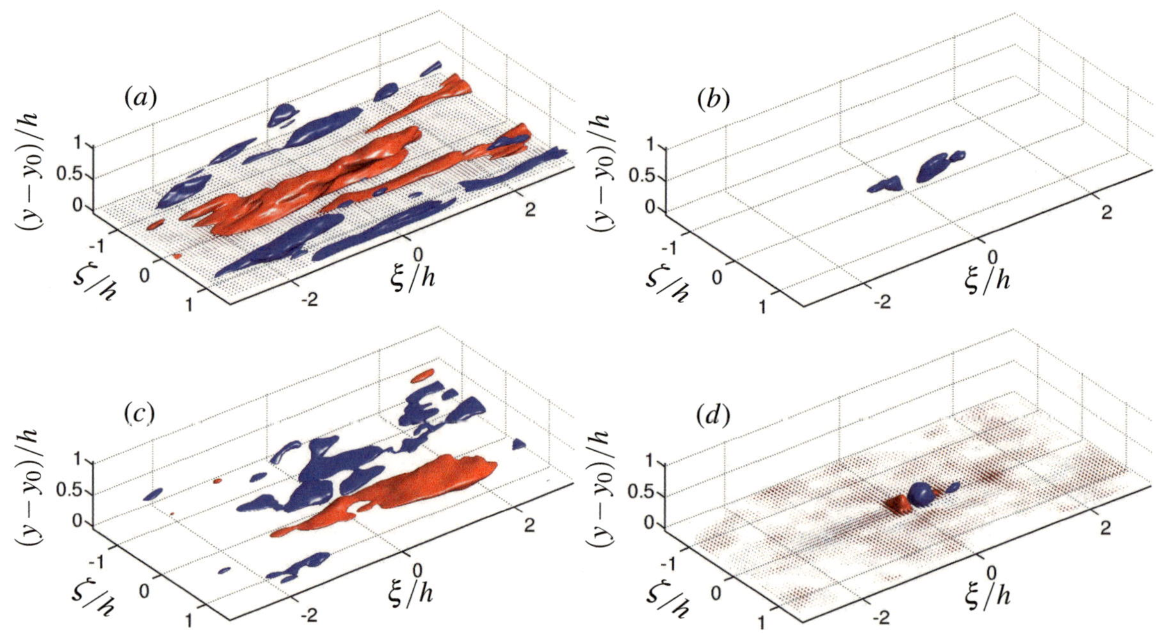

Figure 4.18: Conditionally averaged flow field in case F10 centred around particles with $F_x' > 5\sigma_{Fx}$. Surfaces show iso-surfaces of *(a)* $\langle u' \rangle_{\mathrm{cond}} = \pm 0.8 u_\tau$, *(b)* $\langle v' \rangle_{\mathrm{cond}} = \pm 0.5 u_\tau$, *(c)* $\langle w' \rangle_{\mathrm{cond}} = \pm 0.5 u_\tau$, *(d)* $\langle p' \rangle_{\mathrm{cond}} = \pm 1.5 \rho_f u_\tau^2$ in red (positive) and blue (negative).

number of samples might increase due to the higher intermittency of drag compared to case F50 (cf. table 3.2). Note, that the conditioning criterion as defined by (4.6) does not depend on patterns of drag in time. Thus, two samples of flow fields might be related to each other, e.g. they might stem from the same flow event correlated in time or space. This might occur particularly often in case F10 where the spatial coherence of drag fluctuations is high (cf. figure 4.14). As a result, the increased number of samples in case F10 might not necessarily lead to a better quality of statistical convergence.

Figure 4.18 and figure 4.20 show the conditionally averaged velocity fluctuations in streamwise ($\langle u' \rangle_{\mathrm{cond}}$), wall-normal ($\langle v' \rangle_{\mathrm{cond}}$) and spanwise direction ($\langle w' \rangle_{\mathrm{cond}}$) jointly with the conditionally averaged pressure fluctuation ($\langle p' \rangle_{\mathrm{cond}}$). Figures 4.19 and 4.21 show the respective iso-surfaces in top and side view. The iso-surfaces are found to be not completely smooth which is a strong indication for somewhat marginal convergence of the statistics. A measure of the convergence is the (anti-)symmetry of the surfaces with respect to the ξ-axis which is achieved only approximately in the figures. One reason for the weak convergence is the high threshold criterion chosen, which leads to a small number of samples. Also the values to visualise the conditionally averaged flow fields by iso-surfaces is chosen rather small, i.e. it is smaller than the root-mean-square value close to the wall (cf. figure 3.6). The chosen values are comparable to the values chosen to visualise the correlation between flow field and particle forces in §3.7.2. The resulting structures of the conditionally averaged flow field are large and might be more sensitive to convergence issues.

Most figures show iso-surfaces of alternating value located close to the virtual wall. They cover the entire domain and have sizes similar to the particle diameter. These iso-surfaces are in parts a result from defining flow field fluctuations by the deviation from the plane average as opposed to the deviation from the three-dimensional time-averaged flow field. Thus, these iso-surfaces represent to some degree the three-dimensionality of the mean flow field around the particles (cf. figures 3.9 and

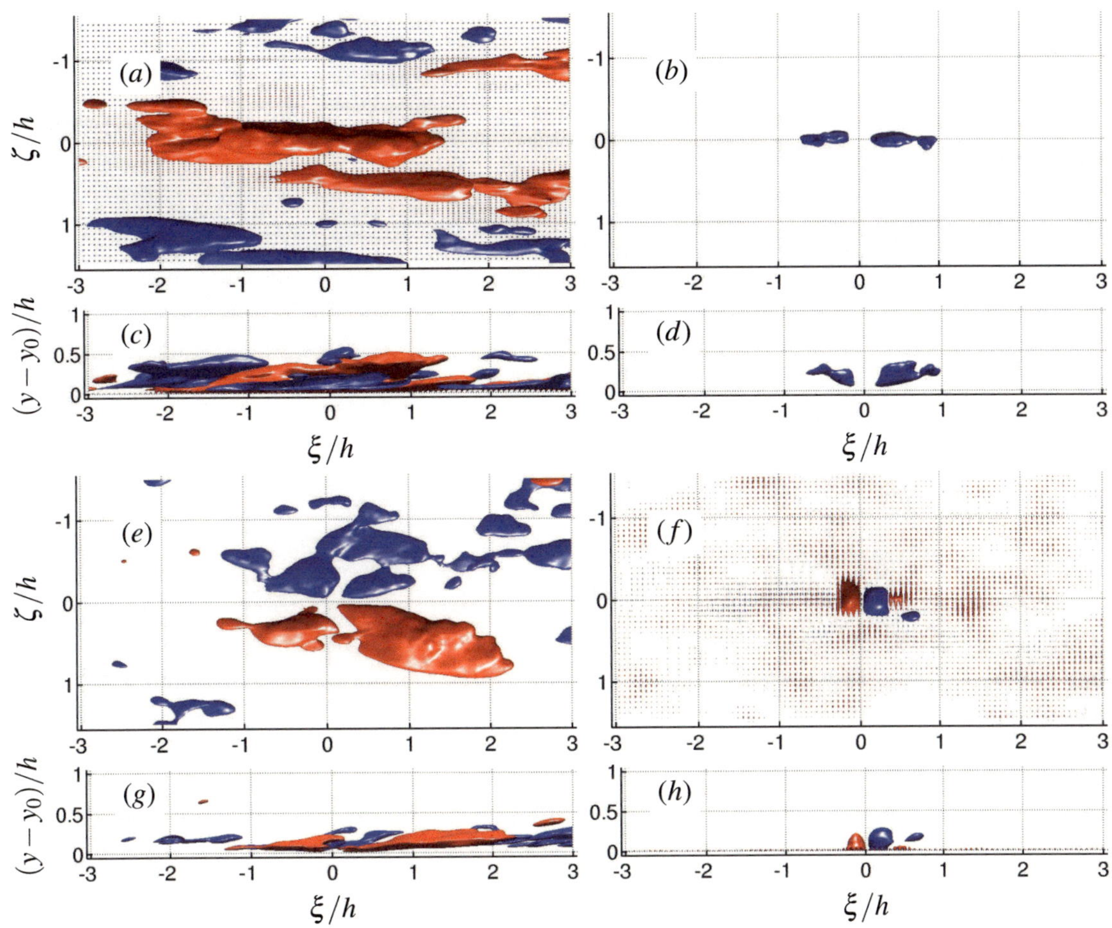

Figure 4.19: Conditionally averaged flow field in case F10 centred around particles with $F_x' > 5\sigma_{Fx}$. Surfaces show iso-surfaces of (*a*,*c*) $\langle u' \rangle_{\text{cond}} = \pm 0.8 u_\tau$, (*b*,*d*) $\langle v' \rangle_{\text{cond}} = \pm 0.5 u_\tau$, (*e*,*g*) $\langle w' \rangle_{\text{cond}} = \pm 0.5 u_\tau$, (*f*,*h*) $\langle p' \rangle_{\text{cond}} = \pm 1.5 \rho_f u_\tau^2$ in red (positive) and blue (negative). (*a*,*b*,*e*,*f*) Top view, (*c*,*d*,*g*,*h*) side view.

3.10). In case F50 these near wall structures can reach considerable heights for the values shown (cf. figure 4.21*h*).

The conditionally averaged streamwise velocity fluctuation, $\langle u' \rangle_{\text{cond}}$, exhibits similar characteristics for case F10 and case F50. In both cases, the iso-surfaces of $\langle u' \rangle_{\text{cond}}/u_\tau = 0.8$ are approximately of size $4h \times 0.5h$ in streamwise and spanwise direction (cf. figures 4.19*c* and 4.21*c*). As before, u_τ is the friction velocity as defined in §3.1 (see also discussion in appendix §C.1). In case F10, the iso-surface of $\langle u' \rangle_{\text{cond}}$ exceeds wall-normal distances of $(y-y_0)/h = 0.5$ while in case F50 even higher values are reached. The iso-surfaces of positive value $\langle u' \rangle_{\text{cond}}$ are flanked by iso-surfaces of negative $\langle u' \rangle_{\text{cond}}$ in both cases. The latter appear to be elongated in streamwise direction similar to the positive iso-surfaces but appear to be less well converged. In case F50, the surfaces seem to be located around $\zeta/h = \pm 1$. In case F10, the spacing appears to be even larger although the quality of the statistics is too low to infer about its trend.

From the analysis in §4.2.1 one might expect $\langle u' \rangle_{\text{cond}}$ to have similar characteristics as the correlation of drag fluctuations to streamwise velocity fluctuations shown in figure 3.38. Indeed, this is found to be true for the present cases. The size of the correlation shown in figure 3.38 compares to size of

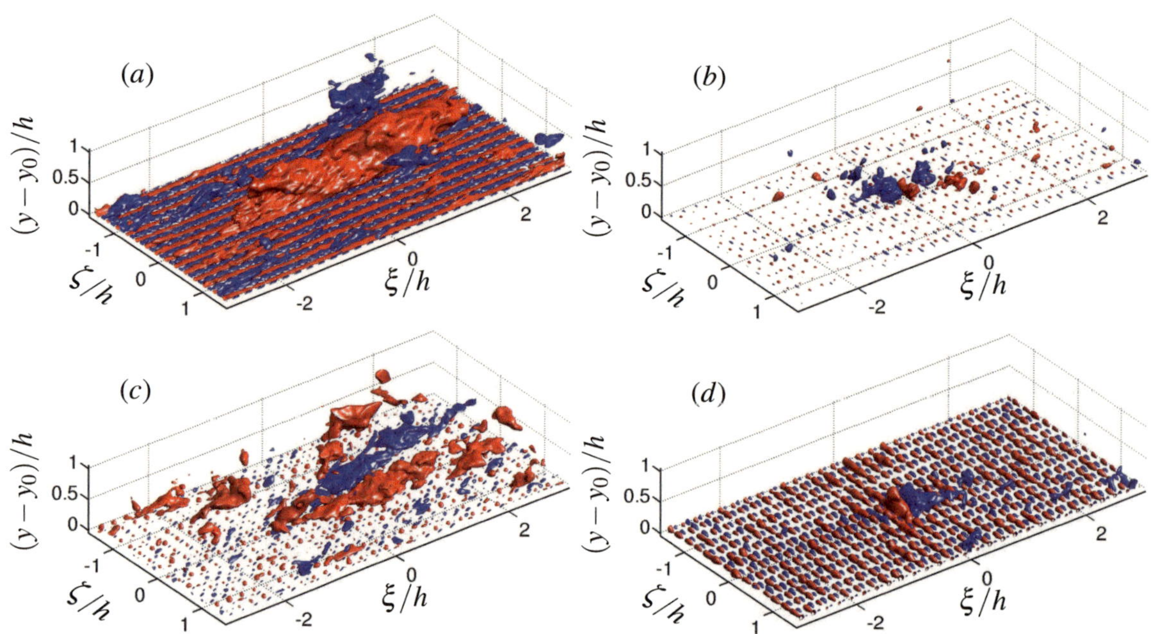

Figure 4.20: Conditionally averaged flow field in case F50 centred around particles with $F'_x > 5\sigma_{Fx}$. Surfaces show iso-surfaces of (*a*) $\langle u'\rangle_{\mathrm{cond}} = \pm 0.8u_\tau$, (*b*) $\langle v'\rangle_{\mathrm{cond}} = \pm 0.5u_\tau$, (*c*) $\langle w'\rangle_{\mathrm{cond}} = \pm 0.5u_\tau$, (*d*) $\langle p'\rangle_{\mathrm{cond}} = \pm 1.5\rho_f u_\tau^2$ in red (positive) and blue (negative).

the corresponding iso-surfaces shown in figures 4.19(*a*) and 4.21(*a*). Recall that this is partly due to the values at which the iso-surfaces are plotted. The iso-value of $\pm 0.8u_\tau$ employed in figures 4.19(*a*) and 4.21(*a*) corresponds to a value of $\pm 0.24\,|\langle u'\rangle_{\mathrm{cond}}|_{\mathrm{max}}$ ($\pm 0.16\,|\langle u'\rangle_{\mathrm{cond}}|_{\mathrm{max}}$) in case F10 (F50). This appears to correspond well to the value of $R/|R|_{\mathrm{max}} = \pm 0.15$ in figure 3.38. However, as R is a very different property as $\langle u'\rangle_{\mathrm{cond}}$ the comparison should be taken with care. Another agreement between the iso-surfaces is the slightly inclined shape of $\langle u'\rangle_{\mathrm{cond}}$. Figures 4.18 and 4.21(*a*,*c*) also reveal a drop of the structure around $\xi/h = 0$ similar to the one found in the correlation F'_x and u' in figure 3.38.

The conditionally averaged wall-normal velocity fluctuation, $\langle v'\rangle_{\mathrm{cond}}$, is shown by iso-surface of $\langle v'\rangle_{\mathrm{cond}}/u_\tau = \pm 0.5$ in figures 4.18(*b*) and 4.20(*b*). In case F10, this corresponds to $\pm 0.54\,|\langle v'\rangle_{\mathrm{cond}}|_{\mathrm{max}}$. Only negative $\langle v'\rangle_{\mathrm{cond}}$ exceed this threshold, positioned upstream and downstream of the particle. In case F50, the value corresponds to $\pm 0.38\,|\langle v'\rangle_{\mathrm{cond}}|_{\mathrm{max}}$. In addition to the negative iso-surfaces positioned upstream and downstream of the particle, positive iso-surfaces appear above the particle. In both cases, the iso-surfaces are approximately of size $2h \times 0.5h \times 0.5$ (figure 4.19*c*,*d* and figure 4.21*c*,*d*). Figures 4.18(*c*) and 4.20(*c*) show the iso-surfaces of $\langle w'\rangle_{\mathrm{cond}}/u_\tau = \pm 0.5$, which corresponds to $\pm 0.45\,|\langle w'\rangle_{\mathrm{cond}}|_{\mathrm{max}}$ ($\pm 0.26\,|\langle w'\rangle_{\mathrm{cond}}|_{\mathrm{max}}$) in case F10 (F50). In both cases, a pair of positive and negative iso-surfaces exists that are approximately confined to the region $-1 \le \xi/h \le 2.5$, $0 \le (y-y_0)/h \le 0.5$ and $-1 \le \zeta/h \le 1$. The sign of the iso-surfaces are such that $\langle w'\rangle_{\mathrm{cond}}$ is directed away from the ξ-axis.

Figures 4.18(*d*) and 4.20(*d*) show the iso-surfaces of $\langle p'\rangle_{\mathrm{cond}}/\rho_f u_\tau^2 = \pm 1.5$, where ρ_f is the fluid density. This corresponds to $\pm 0.16\,|\langle p'\rangle_{\mathrm{cond}}|_{\mathrm{max}}$ ($\pm 0.10\,|\langle p'\rangle_{\mathrm{cond}}|_{\mathrm{max}}$) in case F10 (F50). In both cases iso-surfaces of positive values of $\langle p'\rangle_{\mathrm{cond}}$ occur upstream of the particle, while iso-surfaces of negative values occur downstream of the particle. This is in line with the idea that a negative pressure gradient across the particle will lead to positive drag on a particle. The iso-surface of $\langle p'\rangle_{\mathrm{cond}}$ in case F10

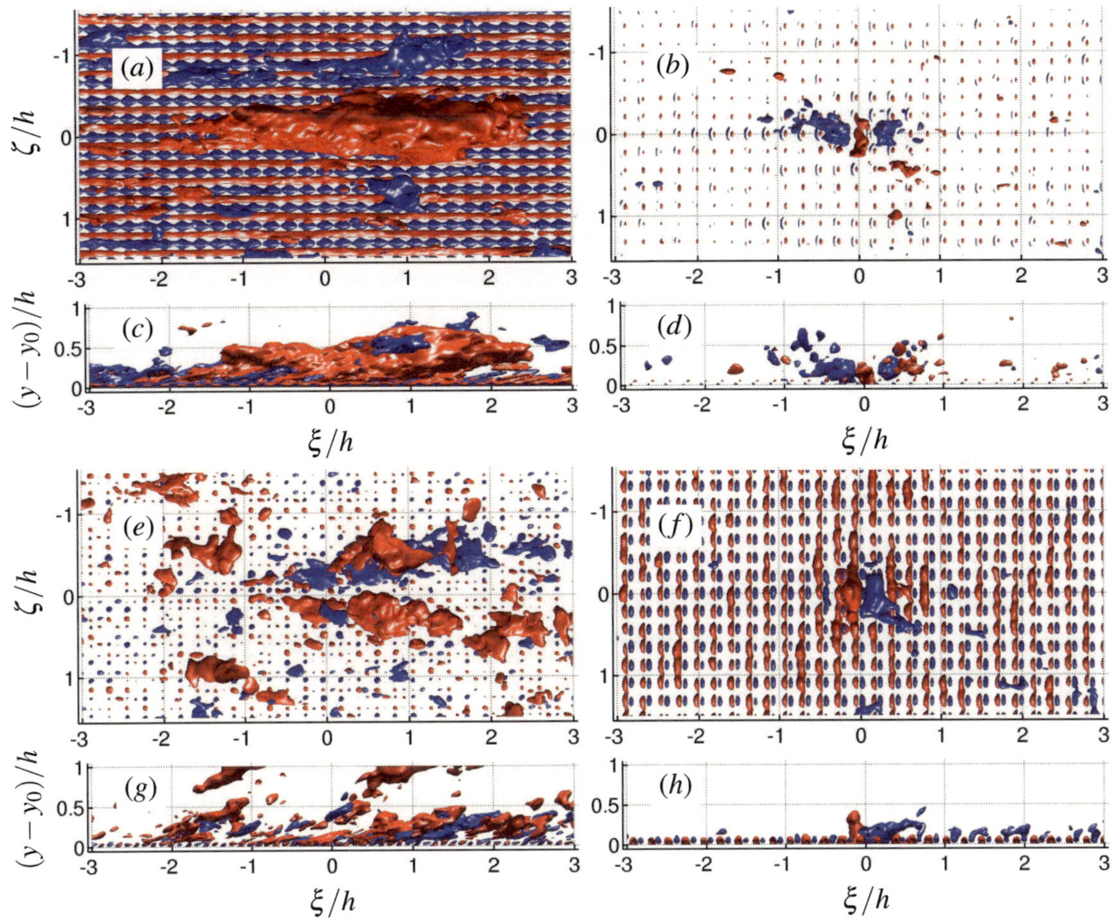

Figure 4.21: Conditionally averaged flow field in case F50 centred around particles with $F_x' > 5\sigma_{Fx}$. Surfaces show iso-surfaces of (*a,c*) $\langle u' \rangle_{\text{cond}} = \pm 0.8 u_\tau$, (*b,d*) $\langle v' \rangle_{\text{cond}} = \pm 0.5 u_\tau$, (*e,g*) $\langle w' \rangle_{\text{cond}} = \pm 0.5 u_\tau$, (*f,h*) $\langle p' \rangle_{\text{cond}} = \pm 1.5 \rho_f u_\tau^2$ in red (positive) and blue (negative). (*a,b,e,f*) Top view, (*c,d,g,h*) side view.

agrees to the correlation in figure 3.42(*a*). As in the correlation, traces of a fork-like extension of the negative pressure region are observed albeit only at one side. A similar extension of the negative iso-surface can be spotted for case F50 in figure 4.20(*d*), despite a similar extension is not visible in the correlation (cf. figure 3.42*b*). However, the convergence of $\langle p' \rangle_{\text{cond}}$ in case F50 appears of lesser quality than in case F10 and the shape of the iso-surface is less well defined.

In line with the present approach is the work of Hofland (2005), Cameron (2006), Detert (2008) and Dwivedi (2010) and the related publications. The authors investigated the flow structures related to force on particle (or its approximation), or, somewhat related, the flow structures related to the onset of sediment erosion. The most promising comparison of the present results is to the results of Detert (2008) and Detert *et al.* (2010*a*). The authors studied flow over a gravel bed at $\text{Re}_b = 1.33 \cdot 10^5$ ($\text{Re}_\tau = 1.26 \cdot 10^4$) with a ratio of $D/h = 19.6$ and a width to effective water depth ratio of $W/h = 4.5$. The gravel are naturally shaped and sieved to obtain a uniform characteristic particle size. The authors present conditionally averaged fluctuations of streamwise, wall-normal, spanwise velocity jointly with conditionally averaged pressure fluctuations on the particle tops. This was possible, due to synchronised velocity and pressure measurements. The conditioning criterion was a high negative gradient of the pressure signal sampled to the maximum peak. This criterion might be interpreted as conditioning the flow field to the maximum related with a sudden increase of lift on the particle. The results are given for two planes at $\zeta/h = 0$ and $(y-y_0)/h \approx 0.04$.

Despite the various differences in the setup of the experiment to the present cases as well as the differences related to the averaging condition, the results compare well. The obtained conditionally averaged flow fields exhibit similar characteristics in both studies. It is remarkable that the results do not only compare in shape but also in strength. The conditionally averaged flow field of Detert *et al.* (2010*a*) reveals positive streamwise velocity fluctuations above the particle, a wall-normal velocity fluctuations towards the wall upstream of the particle as well as wall-normal velocity fluctuations away from the wall. In comparison to the present cases these occur further downstream of the particle. Also in agreement with the results of the simulations, the conditionally averaged spanwise velocity is directed away from the ξ-axis in the vicinity of the particle. However, the conditionally averaged spanwise velocity is more patchy and confined as in the present results.

While the agreement in the conditionally averaged velocities is encouraging, it is surprising to find large differences in the conditionally averaged pressure fluctuations. Similar to here, Detert *et al.* (2010*a*) find two dominant regions of alternating sign located upstream and downstream of the particle. However, the sign of the pressure regions is opposite to here, i.e. it is negative upstream and positive downstream of the particle. This discrepancy should be addressed by future analysis.

4.3 Probabilistic considerations

As has been reviewed in §1.2.3, sediment erosion is often parameterised in terms of the non-dimensional Shields[1] number θ_s as defined by (1.1). Supposing that erosion is initiated by lift forces alone (cf. §4.1.1), the onset of erosion can be characterised by a balance between hydrodynamic lift force, F_y, and buoyant weight of the particle, $F_{y,\mathrm{thres}} = (\rho_p - \rho_f)gD^3\pi/6$, yielding the following expression for the critical Shields parameter:

$$\theta_{\mathrm{thres}} = \frac{2}{3c_L}, \tag{4.7}$$

where the lift coefficient is defined as $c_L = (4F_y)/\left(\pi\rho_f u_\tau^2 D^2\right)$. Here, g is the gravitational constant, ρ_p is the density of the particle and D its diameter. Please note, that in the above definition of c_L a slightly different normalisation than in §3.3 is chosen. It can be seen that for this erosion scenario the critical value of the Shields number is inversely proportional to the lift coefficient at the onset of erosion.

Figure 4.22 shows the probability density function (pdf) of the lift coefficient c_L for the two present cases. To determine the smallest value of the Shields parameter for which sediment erosion can be initiated, the largest occurring value of c_L needs to be considered in each case. Since the pdfs exhibit exponential tails, it is difficult to determine a precise upper bound of c_L. However, when a given (small) minimum probability of observation is fixed, it is clear that the larger spheres (case F50) will yield a larger maximum value of c_L than the smaller spheres (case F10). Consequently, a smaller critical Shields number is obtained for the spheres in case F50.

As was discussed in §3.3 other modes of erosion than the one by pure lift force are possible. One is erosion by sliding motion, as a result of a force, F_t, tangential to the particle surface at the contact point with the downstream neighbouring particles (cf. equation 4.1). A threshold criterion can be defined by considering the buoyancy force on the particle projected onto the direction of the force, i.e.

[1] Albert Frank Shields, American engineer, ⋆ 26 June 1908 † 1 July 1974

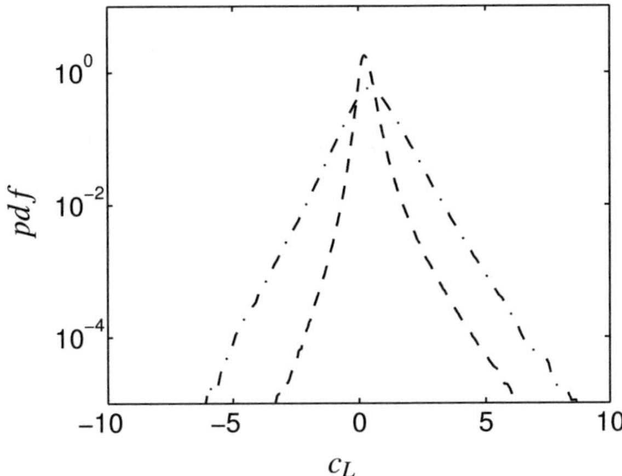

Figure 4.22: Probability density functions of the lift coefficient c_L as defined in text. Lines show case F10 (– – –) and case F50 (– · –).

$F_{t,\text{thres}} = \sin(\alpha)\, g(\rho_p - \rho_f)V_p$, with $\alpha = 35.3°$ and the volume of the particle, V_p. The corresponding Shields parameter in this case might be defined as

$$\theta_{\text{thres}} = \frac{2\sin(\alpha)}{3c_F}, \tag{4.8}$$

where $c_F = (4F_t)\,/\,(\pi\rho_f u_\tau^2 D^2 \pi)$. Another mode of erosion that can be studies is by rotation around the axis through the points of support. The moment, T_E, exerted on the particle with respect to this axis is defined by (4.2). A threshold criterion can be derive from the moment of the gravitational force on the particle with respect to the considered axis of rotation, i.e. $T_{E,\text{thres}} = -r_x^c g\,(\rho_p - \rho_f)\,V_p$, where r_x^c denotes the streamwise distance of the contact points to the centre of the particle. Recall, that the negative torque ($-T_E$) should be considered to be consistent with the notation above. This results in a Shields parameter defined as

$$\theta_{\text{thres}} = \frac{2}{3c_T}, \tag{4.9}$$

where $c_T = -(4T_E)/\left(\pi r_x^c \rho_f u_\tau^2 D^2\right)$. The histogram of c_F and c_T in case F10 and case F50 is given in figure 4.23(*a,b*) respectively. Similar to defining the onset of erosion by lift both modes of erosion indicate a higher critical Shields number in case F10 than in case F50 for a given (small) probability.

Although exhibiting considerable scatter, experiments and field observations seem to indicate an increase with D^+ of the critical Shields number θ_{thres} over the current range of particle diameters (van Rijn, 1993; García, 2008). The different trend found in the present simulations (θ_{thres} decreases from $D^+ = 10$ to $D^+ = 50$) as compared to experiments (θ_{thres} increases from $D^+ \approx 10$ to $D^+ \approx 100$) indicates that extreme force- and torque-generating events recorded in fixed-particle configurations might not directly yield a criterion which is sufficient to judge whether erosion will indeed occur as predicted when particles are freely mobile (under otherwise identical conditions). In particular, the question of the influence of the interaction between the incipient particle motion and the surrounding flow field as well as the influence of possible collective effects during erosion cannot be answered based upon fixed-particle data alone.

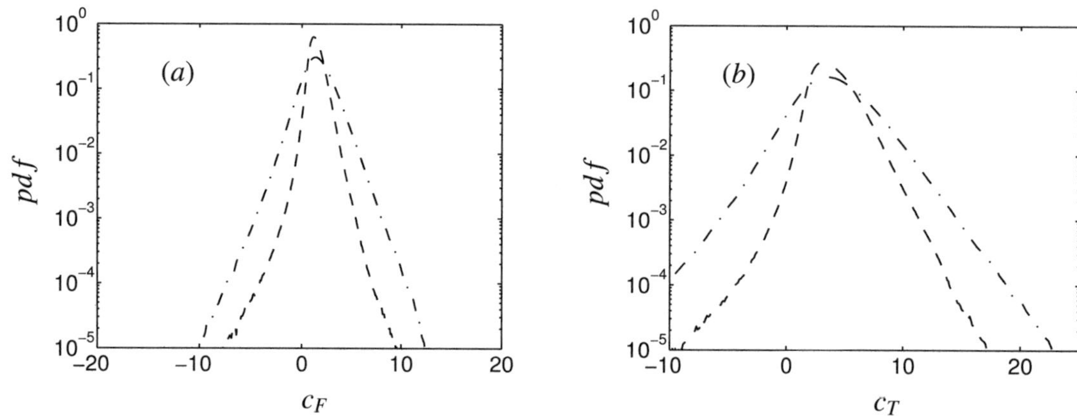

Figure 4.23: Probability density functions of coefficients related to different modes of erosion. (*a*) Pdfs of coefficient c_F of force leading to erosion by translational motion. (*b*) Pdfs of coefficient c_T of negative moment leading to particle erosion rotational motion. Lines show case F10 (– – –) and case F50 (– · –).

4.4 Direct numerical simulation of sediment erosion

This section presents direct numerical simulations of sediment erosion events. The simulation setup is similar to the fixed sphere configuration introduced in chapter 3, however here truly mobile particles are considered that can erode from the bed. The simulations follow the approach of Uhlmann (2006*a*) and are a first step towards deepening the understanding of the mechanism involved in the onset of sediment erosion. However, the small number of studied erosion events in the present study does not allow statistical analysis. Still, the simulations are useful to indicate some trends and guide future research related to direct numerical simulations of the onset of sediment erosion. Some of the results presented here are based on the work of Hannes Strehle during his time at the Institute for Hydromechanics under supervision by the author (Strehle, 2011). His contribution is highly acknowledged. Part of the results, i.e. the analysis of a single erosion event, were previously presented in form of a video-clip (Chan-Braun *et al.*, 2010*c*).

4.4.1 Numerical setup

The setup of the simulations is equivalent to the setup of case F10 and case F50 introduced in chapter 3. In contrast to these fixed sphere simulations, here N_p^f particles are mobile and can erode from the initial bed configuration. To account for the particle–particle contact the artificial repulsion potential of Glowinski *et al.* (1999) (cf. §2.1.2.2) is applied. Note, that a model to account for particle–wall contact is not needed in the present simulations, due to the roughening of the rigid wall below the layer of particles by particle caps (cf. §3.1). The particle caps support the particles in their initial position.

In the following, the simulations of sediment erosion are classified into simulations of multiple particle erosion (M) and simulations of single particle erosion (S). In simulations of multiple particle erosion, all particles in the layer are mobile. In simulations of single particle erosion, only 10 particles are mobile while the other particles remain fixed. The 10 mobile particles have been selected such that they are positioned in regions of high force events during simulations involving only fixed spheres. Furthermore, the mobile particles are located far away from each other to minimise possible collective effects. Figure 4.24 illustrates the position of the released particles in cases S10 by black circles. Additionally, the value of lift prior to releasing the particles is shown by coloured squares.

Case	N_p^f	ρ_p/ρ_f	gh/U_{bh}^2	θ_s	$\tau_c U_{bH}/H$	erosion
S10–L1	10	1.25	-0.69	0.43	6	no
S10–L2	10	1.25	-0.35	0.87	10	no
S10–L3	10	1.25	-0.14	2.17	19	yes
S10–L4	10	1.25	-0.07	4.35	8	yes
S10–L5	10	1.25	-0.03	8.70	8	yes
S10–H1	10	1.7	-0.14	0.78	6	no
S10–H2	10	1.7	-0.07	1.55	6	yes
M10–H1	9216	1.7	-1.38	0.08	13	no
M10–H2	9216	1.7	-0.69	0.16	10	yes
M50–H1	1024	1.7	-0.39	0.11	26	no
M50–H2	1024	1.7	-0.26	0.17	13	yes
M50–H3	1024	1.7	-0.17	0.26	12	yes

Table 4.1: Parameter of simulations to study the onset of single particle (S) and multiple particle (M) erosion. The setup is analogously to that of case F10 (10) and case F50 (50) introduced in chapter 3. However here, N_p^f particles are mobile and able to erode from the initial bed configuration. The ratio of particle density and fluid density is ρ_p/ρ_f, two ratio were considered a low value (L) of $\rho_p/\rho_f = 1.25$ and a somewhat higher value (H) of $\rho_p/\rho_f = 1.7$, gh/U_{bh}^2 is the normalised gravitational volume force θ_s is the Shields parameter defined by (1.1), $\tau_c U_{bH}/H$ is the total observation time of the simulation, the last column indicates whether or not particle erosion occurred.

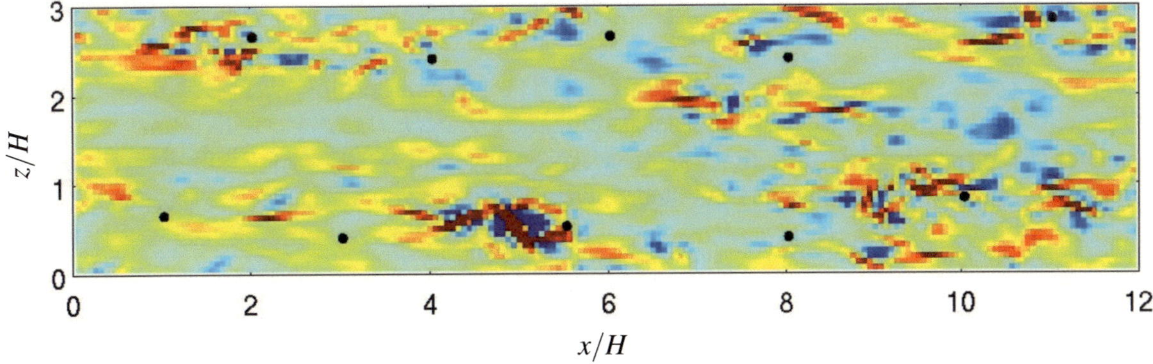

Figure 4.24: Snapshot of case F10 illustrating the position of mobile particles by symbols •. Coloured squares show values of high (low) instantaneous lift force fluctuation F_y'/σ_y^F in red (blue).

Table 4.1 summarises the details of the considered cases. Note that some results differ from Strehle (2011), as some simulations needed to be reconsidered. At the current state of the research, the small sphere case is investigated in most detail. Single as well as multiple particle erosion was studied. Two ratios of particle density to fluid density were considered, one with a low ratio of $\rho_p/\rho_f = 1.25$ (L) and one with a somewhat higher ratio $\rho_p/\rho_f = 1.7$ (H). Both ratios are smaller than the commonly found value of $\rho_p/\rho_f = 2.65$ for rock material in water (for example Vanoni, 1975; van Rijn, 1993) but compare well to previous experiment using Nylon spheres, Ping-Pong balls or artificial stones (Hofland, 2005, $\rho_p/\rho_f = 1.5$, Cameron, 2006, $\rho_p/\rho_f = 1.12 - 1.34$). For the large sphere case only multiple particle erosion events were carried out. In all simulations gravity was applied in direction perpendicular to the wall and successively increased until the particles did not erode. That is, particles

Case	St^+	St_b	Ar
S10–L1	8.05	0.65	266
S10–L2	8.05	0.65	133
S10–L3	8.05	0.65	53
S10–L4	8.05	0.65	27
S10–L5	8.05	0.65	13
S10–H1	10.94	0.89	149
S10–H2	10.94	0.89	75
M10–H1	10.94	0.89	149
M10–H2	10.94	0.89	75
M50–H1	225	11.96	20900
M50–H2	225	11.96	13900
M50–H3	225	11.96	9260

Table 4.2: Stokes numbers, St^+ and St_b, and Archimedes number, Ar, in simulations of sediment erosion.

moved less than half a radius from their initial position during the run-time of the simulations. The initial flow field was taken from case F10 and case F50. A specific conditioning of the instant of time was not needed as local high force events occur evenly distributed over the entire run-time of case F10 and case F50. The duration of each simulation is provided in table 4.1. Note, that the observation times are moderate and longer durations would be preferable. The durations of the simulations are a compromise between computational costs and the minimal time needed for sediment erosion to occur in some cases.

An important parameter in particulate flow is the Stokes number. It is defined as the ratio between a time scale characteristic for the particle, here defined as $\rho_p D^2/(18\rho_f \nu)$, with a time scale characteristic of the flow, here defined either as the viscous time scale, ν/u_τ^2, or the bulk time scale, h/U_{bh}. The two resulting Stokes numbers St^+ and St_b are defined as

$$St^+ = \frac{\rho_p u_\tau^2 D^2}{18\rho_f \nu^2}, \qquad (4.10)$$

$$St_b = \frac{\rho_p U_{bh} D^2}{18\rho_f \nu h}. \qquad (4.11)$$

The value of St^+ and St_b in the present setups are provided in table 4.2. In the small sphere cases that consider a low (high) ratio of ρ_p/ρ_f, the value of St^+ is 8.05 (10.94) and the value of St_b equals 0.65 (0.89). Thus, in the small sphere cases the time scale of the particles is larger than the viscous time scale. However, it is smaller than the time scale of the outer flow. In the large sphere cases the situation is different. Here, St^+ is about 20 times larger than in the respective small sphere case ($St^+ = 225$) and is thus very much larger than viscous time scale. Similarly the value of $St_b = 11.96$ is a factor of 13 larger than in the small sphere case and is therefore much larger than the time scale of the outer flow.

Another non-dimensional number of interest is the Archimedes[1] number, Ar, which is a measure of the effect of the gravitational force with respect to the effect of drag force on a particle (cf. Clift *et al.*, 1978, pp. 113, Uhlmann, 2008, equation 1). Here, the Archimedes number is defined as

$$Ar = \left(\frac{\rho_p}{\rho_f} - 1\right) Re_b^2 \frac{|g|\,h}{U_{bh}^2} \left(\frac{D}{h}\right)^3. \qquad (4.12)$$

[1] Archimedes of Syracuse, Greek mathematician, physicist, engineer and astronomer, ⋆ c. 287 BC † c. 212 BC

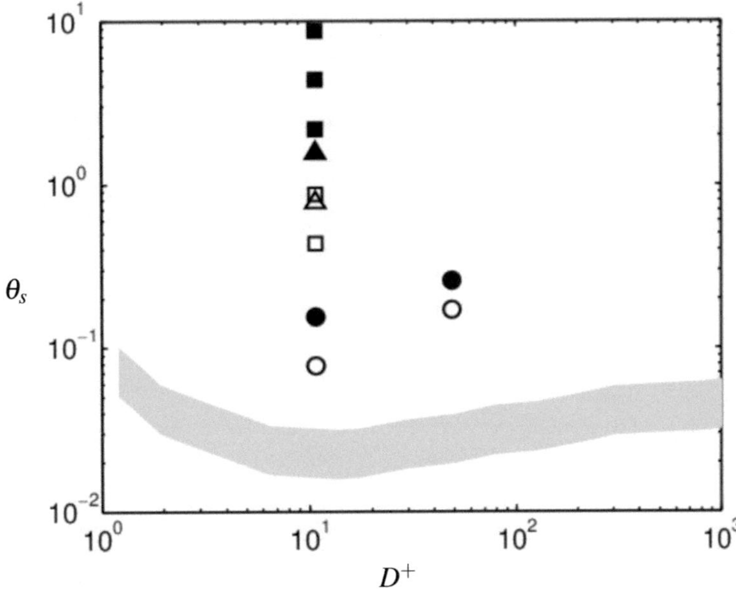

Figure 4.25: Shields parameter θ_s in simulations of table 4.1 as function of particle diameter in viscous units D^+. Empty symbols indicate simulations without particle erosion, bold symbols indicate simulations with particle erosion. Symbols corresponds to cases of multiple particle erosion (○, •), cases S10–L1 to S10–L5 (□, ■), and cases S10–H1 and S10–H2 (△, ▲). Shaded area is range defining the onset of sediment erosion according to Shields (1936).

The value of Ar given in table 4.2 is at least an order of magnitude higher than 1 in all cases considered. The values varies in the range of 13 to 266 for the small spheres case and in the range of 9260 to 20900 in the large sphere case Ar. This indicates that in all cases the motion of the particles is strongly influenced by gravity.

4.4.2 Results

Figure 4.25 illustrates the results of table 4.1. Additionally, the figure shows the area given by Shields (1936) to distinguish the parameter range in which sediment erode (above) and in which particle remain at rest (below). It should be stressed once more, that the results presented here are the results of few simulations. Therefore, the results do not represent mean values and thus do not represent averaged critical Shields numbers. Figure 4.25 indicates that the critical Shields number are higher for the larger sphere case for multiple particle erosion. This is in good agreement with Uhlmann (2006*a*) who carried out similar simulations, albeit in a smaller box and considering particles with sizes of $D^+ = 12.3$ and $D^+ = 108.0$. In the study of Uhlmann (2006*a*) the critical Shields parameter above which sediment erosion occurs was $\theta_{\text{thres}} = 0.12$ in the small sphere case and $\theta_{\text{thres}} = 0.17$ in the large sphere case. These values match well with the present data. The observed trend of an increase in the critical Shields number increases with increasing D^+ from 12 to 108 is in agreement the literature (cf. discussion in §4.3). Figure 4.25 illustrates the trend of the literature by the shaded area provided by Shields (1936). While the trend of the present simulations matches the experimental evidence, the value of the critical Shields numbers are exceptionally high and do not compare with either the results of Shields (1936) with the results of other researchers (cf. Buffington & Montgomery, 1997). This is even more pronounced in case of single sphere erosion, for which the critical Shields number, θ_{thres},

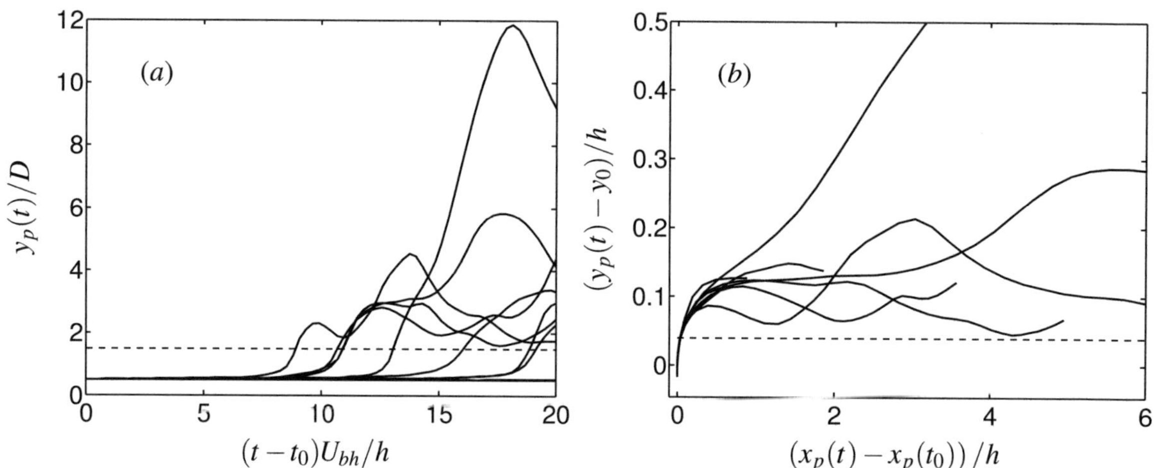

Figure 4.26: Wall-normal position of particle centres, $y_p(t)$, of mobile particles in case S10–L3 (———), (*a*) as a function of time $(t-t_0)U_{bh}/h$, (*b*) as a function of streamwise distance from initial position, $(x_p(t)-x_p(t_0))/h$. Here, t_0 is the initial time of the simulations and $\mathbf{x}_p(t_0) = (x_p(t_0), y_p(t_0), z_p(t_0))$ the initial position of a particle. Dashed lines are introduced to guide the eye and show location of $y = 1.5D$ which is the location of a particle centre positioned on top of a particle with $y_p = D/2$.

is obtained an order of magnitude higher than the values commonly reported. Possible explanation of this shortcoming might be the limitation to short simulations times in the present simulations or the limited number of mobile particles. However, other explanations exist, for example it is well known that the exposure of particles strongly influences the value of the critical Shields number (Fenton & Abbott, 1977; Cameron, 2006). In the present case the eroding particles are initially fully submerged inside of the particle layer. In case of single particle erosion this remains true during the simulation prior of erosion. This aspect might contribute to the obtained higher values of the critical Shields number in particular for the single particle erosion cases.

Figures 4.26 and 4.27 analyse particles quantities during entrainment in simulations S10–L3. Figure 4.26(*a*,*b*) shows the vertical position of the particle centres as a function of time and streamwise distance, respectively. Figure 4.26(*b*) illustrates that most particles in the simulation reach a height of $(y-y_0)/h = 0.12$ within their initial movement. From there they are convected downstream prior to moving back towards the wall or further away from it. However, some particles do not follow this trend. For example the first eroding particle, which erodes around $6h/U_{bh}$ after the start of the simulation, reaches heights of about $(y-y_0)/h = 0.09$, moves back towards the wall after the initial erosion prior to reaching higher values of $(y-y_0)/h$ up to 0.2. Another particle, eroding around $(t-t_0)U_{bh}/h = 13$, is dispersed rapidly far into the outer flow. Here, t_0 is the time at the beginning of the simulations.

Figure 4.27 shows the time signals of the entrained particles with reference to t_{thres} defined as the time for which $y_p(t)$ equals $1.5D$, i.e. $y_p(t_{\text{thres}}) = 1.5D$. Figure 4.27(*a*) shows that the particles reach a wall-normal distance of $y_p/D = 2.2$ within an interval of less than $2h/Ubh$. Remarkable is, that the eroding particles in this simulation are already somewhat lifted from their initial position prior to the entrainment. Figure 4.27(*b*) shows the wall-normal velocity of the particles. The particles are strongly accelerated from about $(t-t_{\text{thres}})U_{bh}/h = -1$ until t_{thres}. The particles wall-normal velocities vary in the range of about $1u_\tau$ to $2u_\tau$. This is in agreement with the results of Cameron (2006) who similarly reported wall-normal particle velocities of the order of u_τ during the entrainment process.

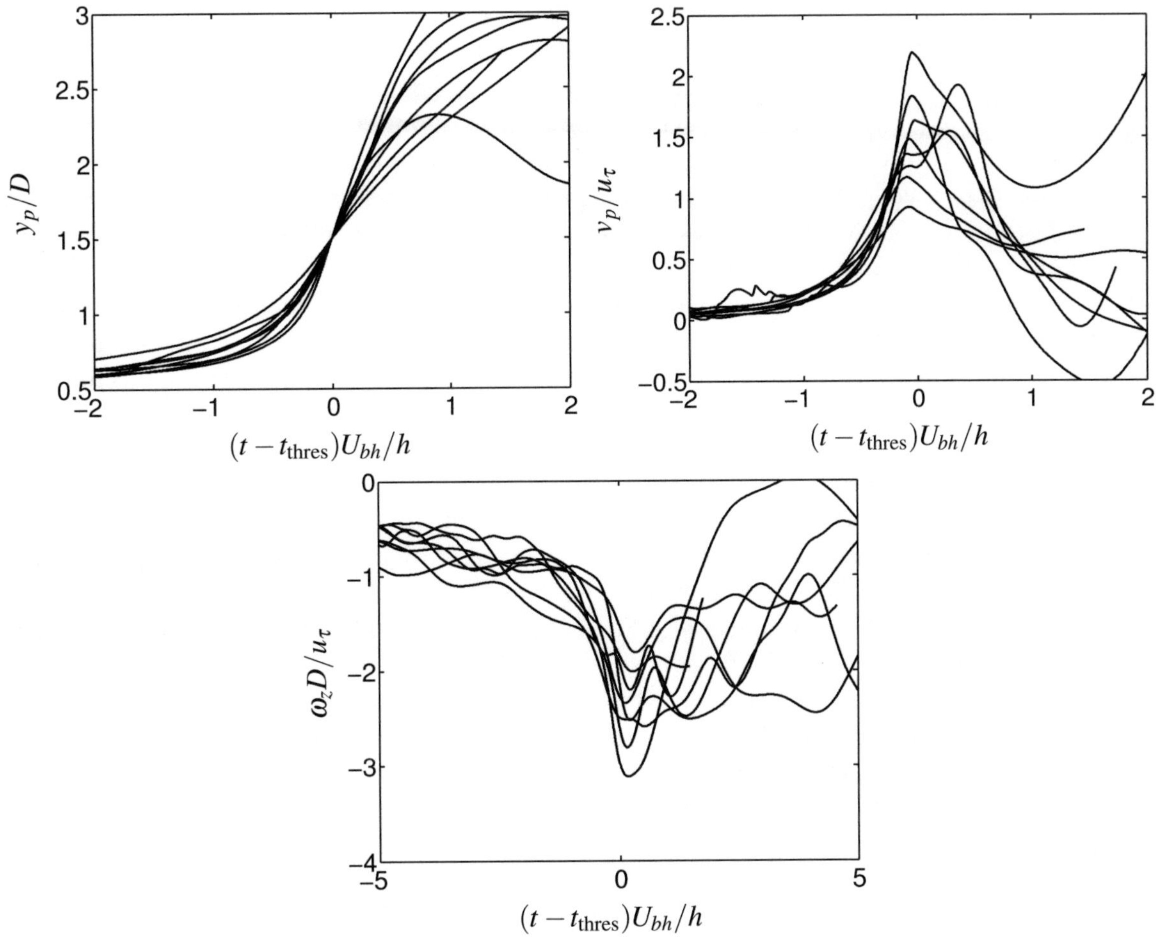

Figure 4.27: Time signals of mobile particles in case S10–L3 (———). (*a*) Wall-normal particle position, y_p/D, (*b*) wall-normal particle velocity, v_p/u_τ and (*c*) spanwise angular velocity, $\omega_z D/u_\tau$. Time signals are shown as a function of $(t - t_{\text{thres}})U_{bh}/h$ where t_{thres} is chosen as the time the respective particle reaches a wall-normal position of $1.5D$, i.e. $y_p(t_{\text{thres}}) = 1.5D$.

These velocities can be regarded as high, i.e. they are higher than the maximum standard deviation of wall-normal velocity close to the wall in case F10 (cf. figure 3.6*a*). Along with the high acceleration in wall-normal direction the particles are exposed to high angular acceleration in spanwise direction (cf. figure 4.27*c*). This could be seen as an indication that high values of lift and spanwise torque – and related to this also drag – coincide in the moment of erosion in the present case. It is remarkable that the wall-normal acceleration of the particle appears to stop as soon as the particle loses its contact from the surrounding particles, i.e. at t_{thres}. It was found that the results of most simulations close the critical Shields number exhibit similar characteristics as those discussed above (plots omitted). In particular this was found to be true not only for the single particle erosion, but also for the cases with multiple particle erosion. While in the small sphere case similar values are obtained, the time of acceleration is about twice as long in the large sphere case. Also in this case the wall-normal velocities of the particles and the wall-normal positions reached are higher.

The following summarises the picture of the erosion process which is derived from analysing the data of the simulations analogously to figure 4.27 as well as with video-clip of the erosion process (cf.

Chan-Braun *et al.*, 2010*c*). In the present simulations, mobile particles start to rotate preferentially along the spanwise axis. Note, that the angular velocity of most particles is non-zero within the time range shown in figure 4.27(*c*). This is caused by the simple contact model employed in the present simulations, that does not consider tangential forces during contact. Such forces would prevent rotation in case of resting particle. In particular in case of multiple particle erosion, mobile particles rotate around the spanwise axis. During erosion particles exhibit a strong acceleration and move away from the wall. This is often related to a duration of about $2h/U_{bh}$ ($4h/U_{bh}$) in the small (large) sphere case. During entrainment the particles are exposed to both high acceleration in wall-normal direction as well as high angular acceleration around the spanwise axis. In case of multiple particle erosion, the motion of particles appears to be strongly correlated along streamwise rows. This collective motion expands in streamwise distances over several h lengths. In the small sphere case the collective motion appears to be smaller than the streamwise length of the domain (Chan-Braun *et al.*, 2010*c*). In the large sphere case the collective motion extents over the entire domain length and entire rows begin to move collectively downstream. A similar phenomenon was reported by Uhlmann & Fröhlich (2007). This among others, indicates that collective effects play an important role for multiple particle erosion in the present cases. Interestingly, the details of sediment erosion as analysed analogously to figure 4.27 are rather similar for multiple as well as single particle erosion.

From the results of Chan-Braun *et al.* (2010*c*) and Strehle (2011) it might be speculated, that for the present small sphere cases single as well as multiple particle erosion might be linked with the passing of a high-speed streak above the eroding particle in combination of a near-by low-speeding streak and hairpin like vortices. The flow structures are fairly similar to the snapshots of flow fields conditioned to high drag in the fixed sphere case presented in figure 4.14. This pictures is in agreement with findings of Hofland (2005) and Cameron (2006) that the entrainment of particle appears to be correlated with events of high streamwise velocity fluctuation jointly with negative wall-normal velocity fluctuations. The authors also reported that the erosion appears to be related with the occurrence of pronounced spanwise vortices. Similar conclusions for the large sphere case would require additional analysis.

4.5 Summary, conclusion and recommendation for future work

This chapter focused on the study of sediment erosion. First the implications of fixed sphere results were considered, then direct numerical simulations with mobile particles were studied.

To investigate events in fixed sphere simulations that could be related to sediment erosion, a criterion is required that predicts sediment erosion on the basis of force and torque on fixed spheres. Such a criterion might be defined by a critical threshold of drag, lift or spanwise torque above which sediment erosion is predicted. More elaborate criteria consider possible motion of particles during erosion based on the given geometry. Here two motions are taken into account, translational motion tangentially to the plane of contact with the supporting particles, and rotational motion around the axis of contact points. The onset of sediment erosion can then be defined by thresholds of the force tangential to the contact plane, or the torque around the axis of support. However, for the present cases it is found that these criteria are equivalent to consider a threshold of drag, i.e. the cross-correlation coefficient of the tangential force with drag is approximately one for the present cases, as is the magnitude of the cross-correlation coefficient of torque around the axis of support and drag. A similar equivalence between the criteria of drag, lift or spanwise torque could not be established for the present cases. The various cross-correlations coefficients between the quantities exceed moderate values. Additionally, the two-dimensional probability density function of drag and lift was examined for time-lags that maximise the correlation coefficient. The results suggest, that although drag and lift are correlated, they might lead to different predictions of the onset of sediment erosion.

The implications of a certain threshold on the conditioning criteria are studied by the number and the mean duration of time intervals. The time intervals are defined as the sequences in which the time series of drag, lift and spanwise force exceed a given threshold. Events of high drag or lift, i.e. with a value of five standard deviation higher than the mean, are rare and on average part of a time sequence with a pronounced single maximum. The mean duration of these sequences is found to be of the order of the smallest scales. In case of spanwise torque, a criterion of five standard deviations above the mean similarly selects rare events with a short mean duration. However, the results of the analysis indicate that even for short durations the signals of spanwise torque contain several local maxima.

Conditionally averaged time signals of drag, lift and spanwise torque on fixed particles are presented. The conditioning criteria was drag, lift and negative spanwise torque five standard deviations higher than the mean. It is found, that the conditionally averaged time signals depend heavily on the conditioning criteria. In particular, the shape of each profile is closely related to the correlation function of the conditioning quantity and the averaged quantity. This appears to be true despite the high threshold employed. The results show, that high drag events correlate to high lift events and vice versa. The conditionally averaged time signals have common features between the small and the large sphere case. In contrast to this, the averaged profiles of spanwise torque conditioned to high drag or lift differ in their characteristics from the small sphere case to the large sphere case. Also, the magnitudes of the profiles are small for the large sphere case. Considering both cases, the highest magnitudes of conditionally averaged drag, lift and spanwise torque are obtained with drag as a conditioning criterion.

Instantaneous flow fields conditioned to high drag events show, that particles of high drag are often located close-by highly negative pressure fluctuations in the vicinity of a pair of high- and low-speed streaks. Frequently, the particles are within a region of high turbulence activity which might resemble burst events. However, in particular for the small sphere case particles can also be in regions where the flow is less active.

Similar to the conditionally averaged time signals of force and torque on a particle, the conditionally averaged flow field compares in shape to the correlation function between flow field and conditioning quantity. A possible explanation of the agreement is to assume, that high drag events dominate the correlation function. However, given the high threshold criterion considered, i.e. five standard deviations higher than the mean, it seems unlikely that these rare force events of small probability should dominate the correlation. Another explanation is to assume, that the flow structures related to drag events are similar in shape independently of the strength of the force and torque. Some support for such an assumption can be gained from the results of Johansson *et al.* (1987). The authors found that pressure signals in a boundary layer conditionally averaged to high and low peaks are similar in shape albeit with opposite sign. This aspect deserves further clarification in future research.

The conditionally averaged flow fields were compared with the results of Detert *et al.* (2010*a*). The authors studied experimentally the conditionally averaged flow over natural gravel at much higher Reynolds numbers. Their conditioning criterion might be interpreted as a strong increase in lift on the particle. Despite the differences in flow configuration and conditioning criterion, the conditionally averaged velocities exhibit similar characteristics in strength and in size. In particular, the authors report high streamwise velocity fluctuations above the particle, elongated in streamwise direction over several particle diameter. Also, the authors observed averaged velocity fluctuations directed towards the wall close-by the particle. The conditionally average spanwise velocities in the vicinity of the particle are directed away from the particle, similar to the results in this study. In contrast to the good agreement of the conditionally averaged velocity fields, the conditionally averaged pressure field in the experiment differs to the present cases. Detert *et al.* (2010*a*) report averaged pressure fluctuations of negative value upstream and of positive value downstream of the particle which, taking the conditionally averaged flow motion into account, appears to be counter intuitive. In the present cases the conditionally averaged pressure fluctuations are of similar size but of opposite sign.

The definition of a Shields number based on instantaneous lift and the gravitational force on fixed particles was used to assess the potential for sediment erosion. From the histograms of lift it was concluded, that the critical Shields number to define the onset of erosion is smaller in the large-sphere case as compared to the small-sphere case. The same trend was found when considering Shields numbers based on the forces projected onto the direction tangential to the downstream contact point between spheres in neighbouring positions as well as when evaluating the balance of angular moments around the contact point. Experiments in the literature with truly mobile particles seem to indicate the opposite trend (increasing critical Shields number with increasing particle size in the range of $D^+ \approx$ 10 to 100). However, these opposite trends do not necessarily imply a contradiction, since additional effects which might play a role in the dynamical process of erosion are not addressed when considering fixed particles. Two important idealisations with respect to real-world sediment erosion are made in the considerations above: the regularity of the geometrical arrangement and the immobility of the particles. Concerning the geometry, it is expected that different particle arrangements, size distributions and shapes will lead to a modification of the forces acting upon sediment particles. In particular, varying the protrusion of individual particles has been shown to have a significant effect on the onset of erosion (Fenton & Abbott, 1977; Cameron, 2006). In this respect, the fixed particle configuration studied in the present work can be considered as a case where mutual sheltering of particles is high due to their uniform diameter, spherical shape and regular arrangement.

Concerning the immobility of the particles, several consequences arise from this idealisation which could potentially affect the implications for sediment erosion: (i) the modification of the flow field by the particles during the incipient motion; (ii) the determination of the temporal duration of force- and torque-generating flow events which is necessary in order to achieve irreversible onset of particle motion; (iii) the influence of collective mobility. In order to evaluate some of these mobility effects, additional simulations of sediment erosion with mobile particles were carried out.

The simulations with mobile particles require a model to account for the particle-particle contact. Here, a simple repulsion force model was applied which only considers contact forces normal to the plane of contact and does not account for tangential friction forces. For both cases erosion events are simulated in which all spheres of the particle layer are mobile. The ratio of particle density to fluid density was 1.7 in both cases. The gravity towards the wall was varied to achieve different Shields numbers. For the few simulations considered the critical Shields number was found to be higher for the large sphere case than for the small sphere case. Thus the simulations are in line with the trend of previous experiments. However, the value of the critical Shields numbers in the simulations are considerably higher than the values in the literature. This could be related to the limitation of the direct numerical simulations to small observation times and moderate numbers of mobile particles. Another explanation could be, that for the present setup sheltering effects might be strong (see discussion above).

Particle erosion with an entire layer of mobile spheres exhibit collective motion of particles. To assess the implications of collective effects on the onset of erosion, additional simulations of particle erosion were carried out in which single, isolated spheres are mobile and erode from an otherwise fixed layer of spheres. These simulations have been considered for the small sphere case only. The critical Shields numbers was much higher in these simulations, which suggests that collective effects indeed play a role in the previous simulations.

The time signals of position, wall-normal velocity as well as spanwise rotation of the eroding particle were found to have similar characteristics in most erosion events. It is found that eroding particles are accelerated strongly during erosion. In agreement with experimental results, particles reach high wall-normal velocities of the order of the shear velocity. For the small sphere case, high-speed streaks are located above the eroding particle jointly with highly negative pressure fluctuations close-by. Thus, in the very few studied events of erosion the flow field is similar to the flow field conditioned to high drag on fixed particles.

While the presented studies were helpful to discuss some trends and characteristics of erosion events, the results are preliminary and report work in progress. Additional analyses including supplementary simulations are needed to clarify some open questions. Simulations of single sphere erosion for the large sphere case could clarify whether the prediction of smaller critical Shields numbers, based on the fixed sphere results, stems from neglecting collective effects. A strong simplification in the present simulations was done by neglecting the effect of friction in the particle contact model. This causes mobile particles to rotate with respect to the spanwise axis, even when remaining at their initial position. It is reasonable to assume that this is different for particle made of different materials. Simulations with a more elaborated contact treatment which models the effect of friction would be desirable. Furthermore, in order to quantify the observations of the erosion process by statistical measures many more sediment erosion events are necessary.

Chapter 5

Sediment transport in open channel flow

This chapter focuses on open channel flow with sediment transport of many mobile particles. It provides a discussion on the statistics of flow field and particle related quantities obtained through direct numerical simulations. Two setups with different mass loadings were studied with each a coarse and a fine resolution of the flow field. The analysis and discussion of this chapter follows the one presented in Chan-Braun *et al.* (2010*a*) and Chan-Braun *et al.* (2010*b*). It should be noted that the simulations are similar to those carried out by Kidanemariam (2010). However, in the present work the wall is geometrically rough, whereas it is smooth in the latter reference.

In section 5.1 the setup of the simulations is described and basic definition of parameters are provided. The results of the simulations are discussed in section §5.2, the instantaneous flow field is presented in §5.2.1, a note on statistical aspects is given in §5.2.2. In section 5.2.3 the statistics of the particles and the fluid are analysed. The implications for sediment transport with respect to approximations of the density profiles in rivers are addressed in §5.2.4. The chapter ends with a summary and recommendation for future work in §5.3.

5.1 Numerical setup

The flow configuration consists of turbulent open channel flow at a bulk Reynolds number $\mathrm{Re}_b = U_{bH}H/\nu = 2900$ with suspended mobile spherical particles over a rough wall. As before, ν denotes the kinematic viscosity of the fluid and U_{bH} is the bulk velocity based on the domain height, H, i.e. $U_{bH} = 1/H \int_0^H \langle u \rangle \,\mathrm{d}y$. Details on the flow configuration and numerical setup are given in table 5.1. Two different setups are studied, one with a global solid volume fraction of ϕ_s^g=0.311% and a total of 2000 mobile spheres, another with a global solid volume fraction of ϕ_s^g=1.38% and a total of 9216 mobile spheres. Here, the solid volume fraction is defined as $\phi_s^g = N_p^s \pi D^3/(6L_xL_yL_z)$. For each case two simulations have been carried out at a high and low resolution. The particles are mono-sized and have a ratio of particle diameter, D, to domain height, H, of $D/H = 12/256$. The ratio of particle density, ρ_p, to fluid density, ρ_f, is ρ_p/ρ_f=1.7. A gravitational volume force, g, is applied in wall-normal direction as given in table 5.1. Similar to the setup of case F10 in chapter 3, the wall in the simulations is formed by a layer of fixed spheres in a square arrangement (cf. grey spheres in figure 5.1). The rough wall consists of 192 particles in streamwise and 48 particles in spanwise direction and a distance of $1/16H$ between the particle centres. At $y = 0$ a rigid wall is located below the layer of spheres which is additionally roughened by spherical caps (cf. figure 5.1). The caps can be defined as the part above $y = 0$ of spheres located at $-0.357D$ staggered in the streamwise and spanwise direction with respect to the layer of spheres above. The flow is assumed to be periodic in streamwise

Case	U_{bh}/u_τ	Re_b	Re_τ	D^+	D/Δ_x	Δ_x^+	N_p^s	ϕ_s^g	gh/U_{bh}^2	$\tau_c U_{bH}/H$
T03C	14.4	2880	199	9.7	6	1.62	2000	0.0030	-0.702	273
T03	13.8	2870	207	10.1	12	0.84	2000	0.0030	-0.706	113
T14C	13.7	2880	209	10.2	6	1.70	9216	0.0138	-0.702	284
T14	13.1	2870	218	10.6	12	0.89	9216	0.0138	-0.706	280

Table 5.1: Setup parameters of simulations with sediment transport, U_{bH} is the bulk velocity based on the domain height, H, U_{bh} is the bulk velocity based on the effective open channel height, h, defined as $h = H - 0.8D$, u_τ is the friction velocity, $\mathrm{Re}_b = U_{bH}H/\nu$ is the bulk Reynolds number with ν as the kinematic viscosity, $\mathrm{Re}_\tau = u_\tau h/\nu$ is the friction Reynolds number, $D^+ = Du_\tau/\nu$ is the particle diameter in viscous units, D/Δ_x is the resolution of a particle, $\Delta_x{}^+$ is the grid spacing in viscous units, N_p^s is the number of mobile particles, $\phi_s^g = N_p^s \pi D^3/(6L_xL_yL_z)$ is the global solid volume fraction, g is the gravitational volume force in wall-normal direction, τ_c is the time over which statistics were collected. In all simulation the ratio of ρ_p/ρ_f equals 1.7.

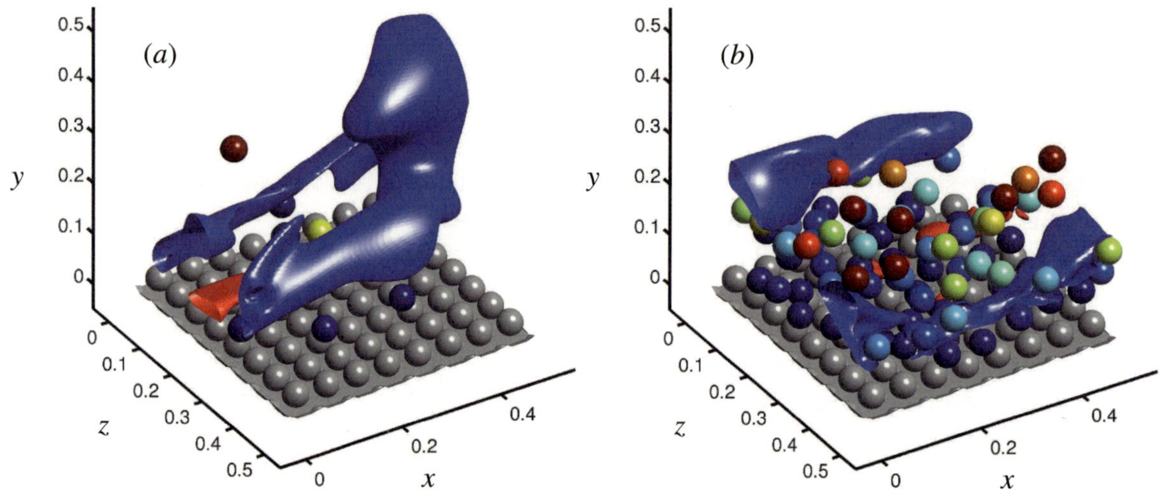

Figure 5.1: Close-up of a section of the computational domain showing the geometry of the bottom wall consisting of a layer of fixed spheres arranged on a square lattice (grey) as well as particle of the solid phase (colour according to height from blue to red). Additionally, iso-contours of the streamwise velocity fluctuation are shown at values $+3u_\tau$ ($-3u_\tau$) in red (blue). Panels show case T03 (*a*) and case T14 (*b*).

and spanwise direction. At the upper boundary a rigid-lid assumption is employed. At the bottom boundary and on the surface of the fixed spheres a no-slip condition is applied.

The computational domain dimensions are $L_x/H \times L_y/H \times L_z/H = 12 \times 1 \times 3$, in streamwise, wall-normal and spanwise directions, respectively. In case T03 and T14 an equidistant Cartesian grid was employed throughout the entire domain using $3072 \times 256 \times 768$ grid points. This correspond to a resolution of the flow field of less than a wall-unit in each coordinate direction and a resolution of the particles of 12 grid point in each coordinate direction. Such a resolution can be defined as very fine away from boundaries and fine close to boundaries. In case T03C and T14C the resolution is coarser by a factor of two and, while the resolution of the flow field far away from the wall can be considered very fine, the resolution near the wall is marginal. Also the resolution of particles with only 6 grid points per diameter in each coordinate direction has to be considered as coarse, the more as the particles are discretised by an immersed boundary method employing a stencil width of $3\Delta_x$. The immersed boundary method as well as the numerical method used in the current simulation is

Case	T03C	T03	T14C	T14
St^+	8.9	9.6	9.8	10.7
St_b	0.644	0.643	0.644	0.643

Table 5.2: Stokes number based on viscous scales, St^+, and Stokes number based on bulk scales St_b in simulations with sediment transport. In all simulations the Archimedes number, Ar, is 470.

discussed in more detail in §2.2.3 and the references therein. The particle–particle contact as well as particle–wall contact has been modelled by a simple repulsion force as discussed in section 2.1.2.2. Recall, that this contact model does not consider tangential forces during contact. Note, that particle–wall contact was applied on the rigid-lid boundary to avoid particles leaving the domain.

The position of the virtual wall, y_0, and the friction velocity, u_τ, are defined as in chapter 3, i.e. y_0 is defined as $y_0/D = 0.8$ and u_τ is defined by extrapolation of the Reynolds stress of the fluid, $\langle u'v' \rangle$, to y_0. A detailed discussion on the subject related to single-phase flow over fixed spheres is given in appendix §C.1. Based on the definition of u_τ and y_0 other quantities are defined analogously as in chapter 3, i.e. the effective channel height, $h = H - y_0$ and the bulk velocity based on the effective channel height, $U_{bh} = 1/h \int_{y_0}^{H} \langle u \rangle \,\mathrm{d}y$.

In the simulations the flow was driven by a time-varying volume force such that the bulk Reynolds number of the flow, $\mathrm{Re}_b = U_{bh}h/\nu$, is constant (cf. table 5.1), where ν is the kinematic viscosity of the fluid. This bulk Reynolds number corresponds to a friction Reynolds number of $\mathrm{Re}_\tau = u_\tau h/\nu = 182$ in case of a smooth wall and to $\mathrm{Re}_\tau = 188$ in case of a single-phase flow over fixed spheres in a similar setup (cf. case F10 in chapter 3). In the present two-phase flow, the friction Reynolds number increases to $\mathrm{Re}_\tau = 207$ in case T03 and to $\mathrm{Re}_\tau = 218$ in case T14. With the increase in Re_τ the particle diameter in viscous units, $D^+ = Du_\tau/\nu$, increases from 10.1 in case T03 to 10.6 in case T14 (cf. table 5.1). It is interesting to note, that the values of Re_τ and D^+ are about 4% lower in the respective coarse simulations.

The values of St^+ as defined by (4.10) vary between 9 and 11 in the simulations, indicating that the time scale of the particles is larger than the viscous time scale. The value of St_b as defined by (4.11) equals 0.64 in each case, showing that the time scale of the particles is smaller than the time scale of the bulk flow. The Archimedes number as defined by (4.12) equals 470 in all cases.

After the flow reached a statistically stationary state, the simulations were continued for over $270H/U_{bH}$, in case T03C, T14C and T14 and continued for $113H/U_{bH}$ in case T03 (cf. table 5.1). Statistics of the flow field as well as particle related data were collected during run-time of the simulations. Snapshots of the flow field jointly with the corresponding particle data was stored in sequences of about $2H/U_{bH}$.

5.2 Results and discussion

5.2.1 Instantaneous flow field

Figures 5.2 and 5.3 show snapshots of the flow field illustrating the complexity of the phenomena involved. The flow field is visualised by iso-surfaces of positive (negative) streamwise velocity fluctuations in red (blue). Moving particles are coloured according to their distance to the wall. The figures show that the particles are small with respect to the domain size and the flow structures in both cases. Due to the gravitation force in wall-normal direction particles tend to accumulate close to the wall. Occasionally particles reach the upper part of the domain as a result of the turbulence–particle

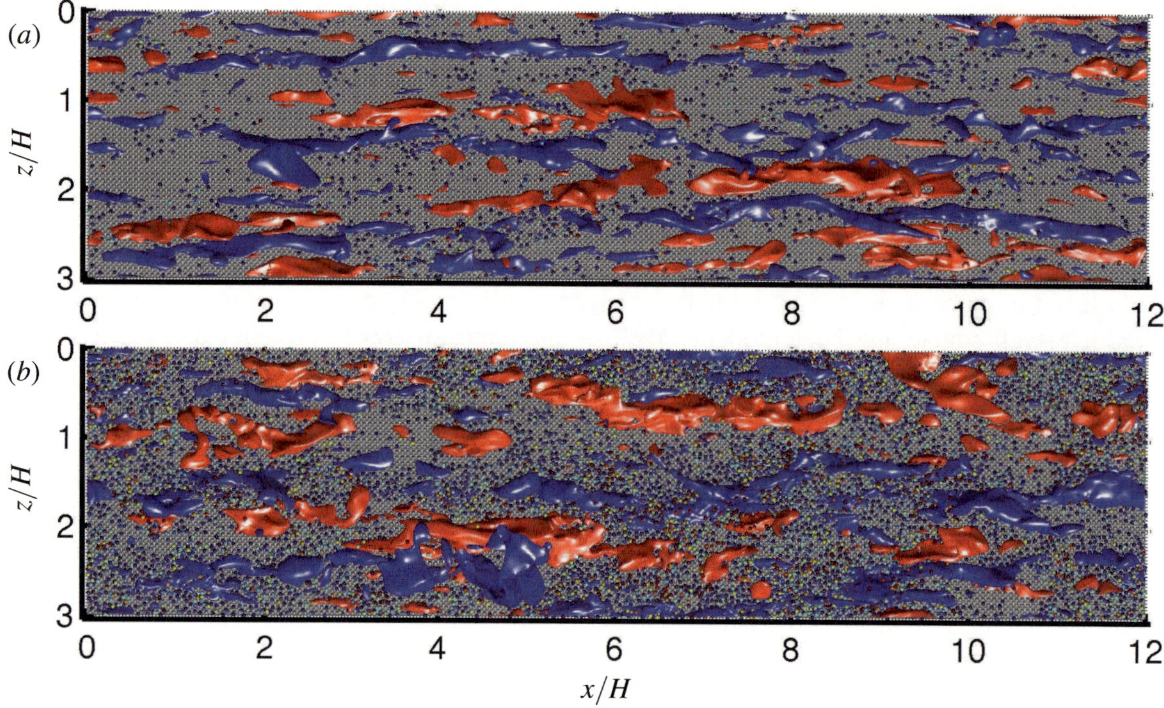

Figure 5.2: Top view of instantaneous flow field in case T03 (*a*) and case T14 (*b*). Red (blue) surfaces are iso-contours of the streamwise velocity fluctuation at values $+3u_\tau$ ($-3u_\tau$). Mobile particles are coloured according to their height from blue to red.

interaction. The difference in the solid volume function is clearly visible in the figures, while in case T03 the number of mobile particles is moderate it increases considerably in case T14. Figures 5.2 and 5.3 reveal modifications of the turbulence structures in comparison to those of smooth (figure B.1) and rough wall simulations (figures 3.2 and 3.3).

5.2.2 Note on statistical analysis

In the following sections Eulerian[1] statistics of the flow field and of the particles as a function of wall-distance are discussed. The flow field statistics are collected during run-time and are computed over the fluid phase only (cf. equation C.3). Eulerian statistics of particle related data are computed from discrete bins in wall-normal direction analogously to Uhlmann (2008). The details on the definition of the number density, n_s, the solid volume fraction, ϕ_s, and Eulerian statistics of a quantity ϕ_p related to moving particle are provided in appendix D. Note, that in the present study the bins are stretched in wall-normal direction by a factor of $\alpha_{\mathrm{bin}} = 1.05$ in order to reach a fine resolution for the steep gradients close to the wall and a higher number of samples in bins far away from the wall. Here, 100 bins are considered in wall-normal direction.

Once a statistically stationary state is reached, the concentration of particles is much higher close to the wall than in the upper part of the channel due to gravity. This can be seen in figure 5.4 that shows the distribution of the mean solid volume fraction, $\langle\phi_s\rangle$, as defined by (D.5). The solid volume fraction and thus the time and plane averaged particle number density, $\langle n_s\rangle$, defined by D.4 decreases

[1]Leonard Euler, Swiss mathematician, $\star$ 15 April 1707 † 18 September 1783

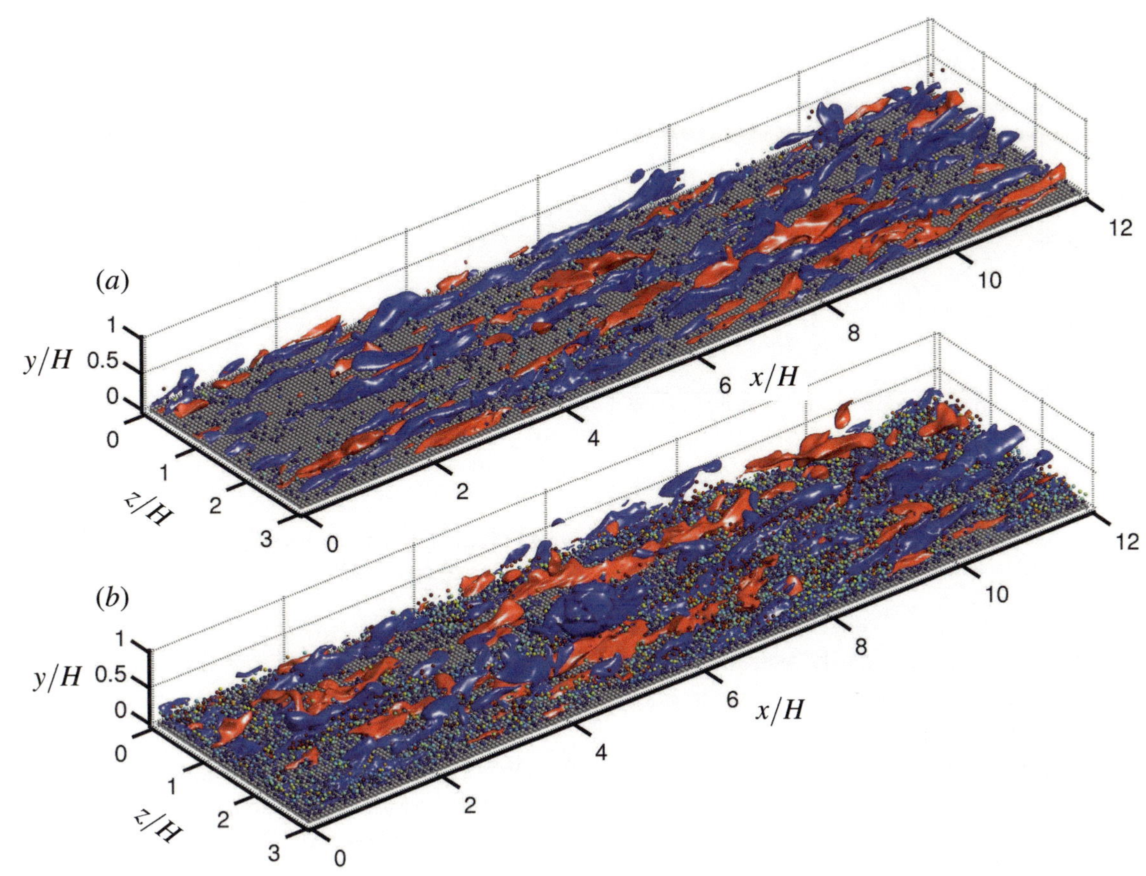

Figure 5.3: Instantaneous flow field in case T03 (*a*) and case T14 (*b*). Red (blue) surfaces are iso-contours of the streamwise velocity fluctuation at values $+3u_\tau$ $(-3u_\tau)$. Mobile particles are coloured according to their height from blue to red.

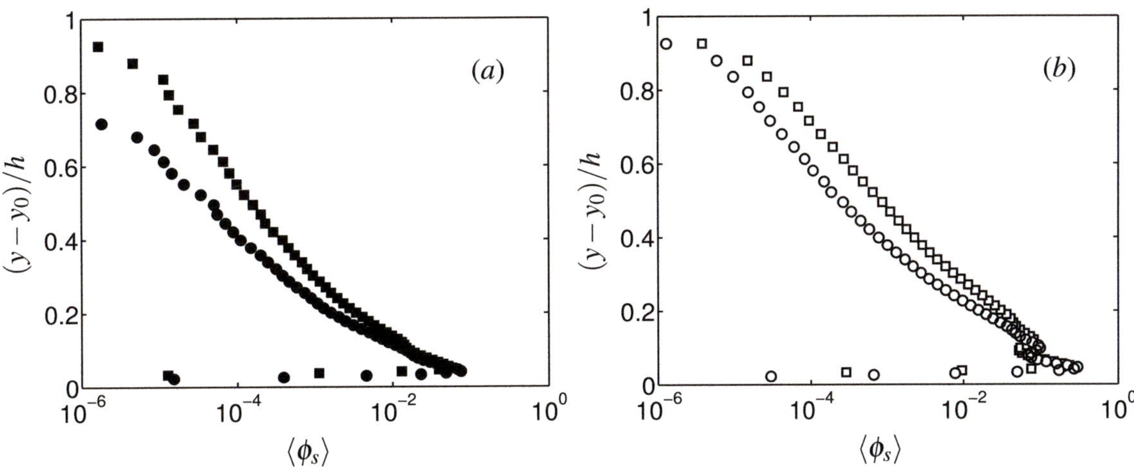

Figure 5.4: Mean solid volume fraction, $\langle\phi_s\rangle$, as function of $(y-y_0)/h$. (*a*) Case T03C (■) and case T03 (•), (*b*) case T14C (□) and case T14 (○).

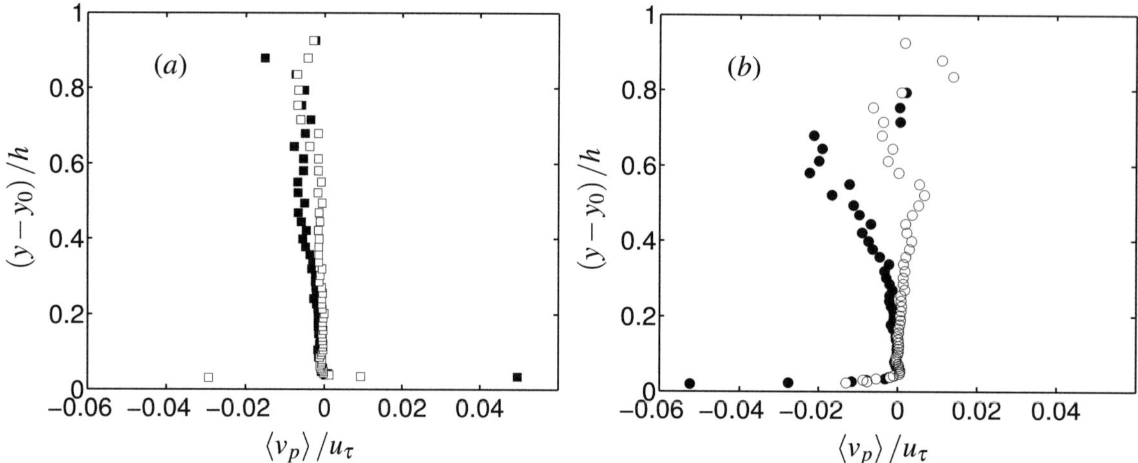

Figure 5.5: Mean wall-normal velocity component of solid phase, $\langle v_p \rangle$ normalised by friction velocity, u_τ as a function of $(y-y_0)/h$. (*a*) Case T03C (■) and case T14C (□), (*b*) case T03 (•) and case T14 (○).

approximately exponentially over most of the flow depth. Thus, in the upper part of the domain only few samples are collected in the observation interval and fully converged particle statistics over the entire flow depth are hard to obtain.

As a measure of statistical convergence of the dispersed flow statistics figure 5.5 displays the wall-normal profile of the mean wall-normal velocity of the particles, $\langle v_p \rangle$. In case of fully converged statistics $\langle v_p \rangle$ should be zero for all y. The largest deviation from the zero value in the outer flow are of the order of $0.02u_\tau$. This value is rather small and the simulations can be regarded as statistically stationary. For $(y-y_0)/h < 0.3$ the error is smaller except close to the wall, where it increases due to lack of samples in the near wall bins (cf. figure 5.4). In case T03C and T14C, the statistics of the mean appear to be reasonable well converged over most part of the flow depth, while in particular in case T03 the statistics appear to be converged only for $(y-y_0)/h < 0.4$. Above $(y-y_0)/h \approx 0.4$ the scatter of the data increases. As has been explained above, the latter region is visited by few particles. In order to obtain well-converged particle statistics in regions far from the wall considerably longer observation times would be required.

5.2.3 Flow field statistics

The results of the two-phase simulations are compared with the results of a single-phase open channel flow over a smooth bed at the same bulk Reynolds number $\mathrm{Re}_b = 2880$, which corresponds to a friction Reynolds number of $\mathrm{Re}_\tau = 183$. The single-phase simulation was performed with the in-house code LESOCC2 (Breuer & Rodi, 1996; Hinterberger, 2004). In the following this simulation will be referred to as smooth wall simulation for convenience. Details on the simulations with a comparison to similar single-phase open channel flows can be found in appendix B.

The mean streamwise velocity profiles of fluid phase and disperse phase in the simulations are given in figures 5.6 and 5.7 in comparison with the result of the smooth wall simulation. First focus is given on the results of simulations T03 and T14. The effect of the disperse phase on the profiles of the fluid phase is comparable to the effect of roughness, i.e. the velocity profiles in figure 5.7(*a*) flattens as a result of the mass loading and in figure 5.7(*b*) the velocity profiles shift towards lower values of $\langle u \rangle^+$. The effect increases somewhat with the global volume fraction, e.g. from simulation T03

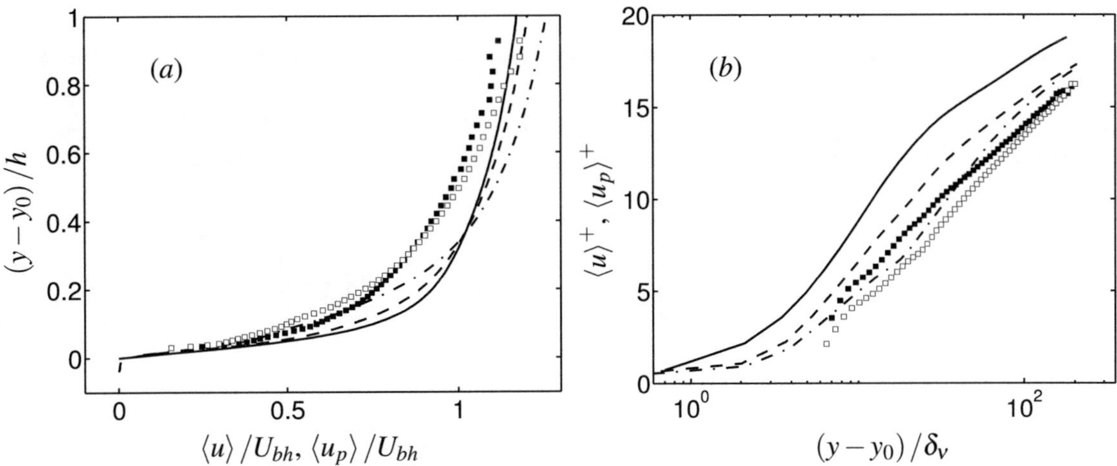

Figure 5.6: Mean streamwise velocity component of fluid phase (lines), $\langle u\rangle$, and solid phase (symbols), $\langle u_p\rangle$, in case T03C and T14C in comparison with results of a single-phase open channel flow over a smooth wall (———). (*a*) Profiles normalised with U_{bh} as a function of $(y-y_0)/h$, (*b*) profiles in semi-logarithmic scale normalised by δ_v and u_τ. Lines and symbols show results of case T03C (– – –, ■), and case T14C (– · –, □)

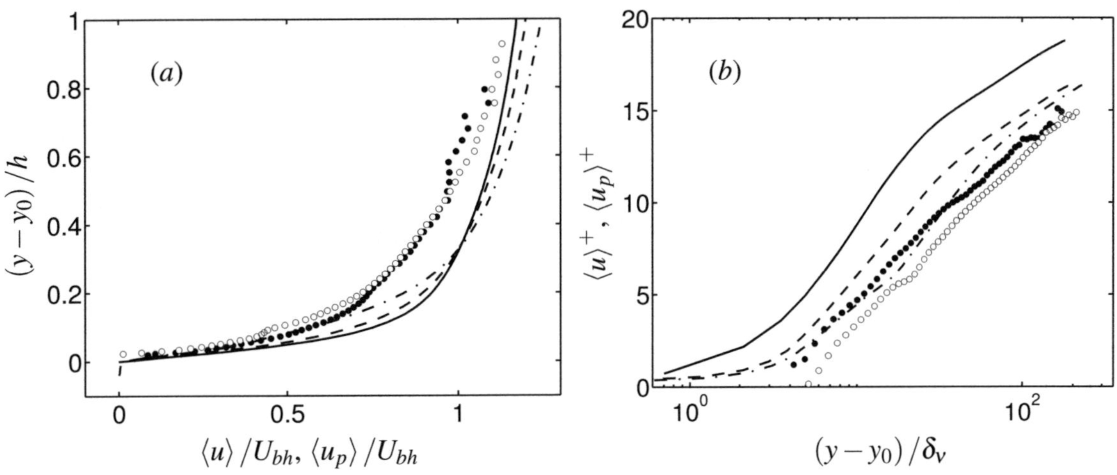

Figure 5.7: Mean streamwise velocity component of fluid phase (lines), $\langle u\rangle$, and solid phase (symbols), $\langle u_p\rangle$, in case T03 and T14 in comparison with results of a single-phase open channel flow over a single-phase open channel flow over a smooth wall (———). (*a*) Profiles normalised with U_{bh} as a function of $(y-y_0)/h$, (*b*) profiles in semi-logarithmic scale normalised by δ_v and u_τ. Lines and symbols show results of case T03 (– – –, •), and case T14 (– · –, ○)

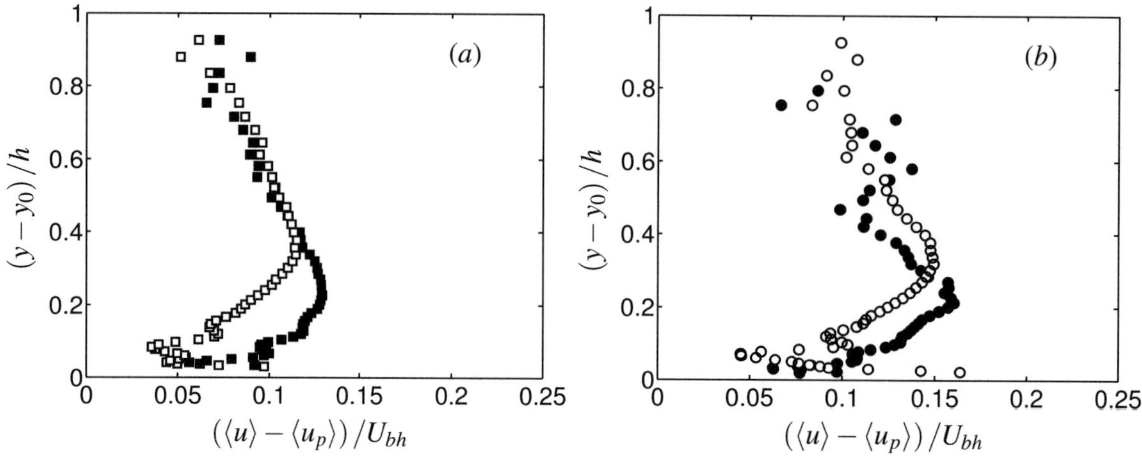

Figure 5.8: Mean streamwise velocity difference between fluid phase and solid phase normalised by bulk velocity as a function of wall distance. (*a*) Case T03C (■) and T14C (□), (*b*) case T03 (•) and T14 (○).

to T14. Additionally, in the high mass loading case T14, the gradient of the velocity appears to be affected close to the wall. In particular a gradient similar to the smooth wall reference case is obtained only for $(y - y_0)/\delta_v > 90$. Such an effect is not known from roughness, in particular the results of case F50 in chapter 3 did not exhibit a change in the gradient, although in this case the roughness effect is stronger than in case T14. A change of the gradient is also in contrast to the observations in previous experiments of two-phase flow in a horizontal channel in a similar flow regime (Kaftori *et al.*, 1995; Kiger & Pan, 2002). The effect might be caused by the low Reynolds number considered, the moderate ratio h/D used in the present simulation and/or the large concentration of particles in the near-wall region. The comparison of the results in figure 5.6 with those in figure 5.7 shows the fluid phase statistics in cases T03C and T14C differ little to those of cases T03 and T14 discussed above.

It is found, that the mean velocity of solid phase lags behind the fluid phase in both cases (figure 5.7). This aspect is further investigated in figure 5.8, which reveals a positive velocity difference between the phases $(\langle u \rangle - \langle u_p \rangle)/U_{bh}$ over the entire channel depth. Averaging $(\langle u \rangle - \langle u_p \rangle)/U_{bh}$ over the wall-normal direction leads to values of $0.095U_{bh}$ ($0.103U_{bh}$) in case T03 (T14) and $0.089U_{bh}$ ($0.079U_{bh}$) in case T03C (T14C). Note, that the higher values obtained in simulations T03 and T14 might be biased by the less converged results in the upper part of the domain as visible in figure 5.8. The profiles of $(\langle u \rangle - \langle u_p \rangle)/U_{bh}$ in figure 5.8(*b*) exhibit a maximum at $(y - y_0)/h = 0.22$ (0.32) with a peak value of $\phi_s = 0.16$ (0.15) in case T03 (T14). From the maximum the velocity decreases towards a value scattered around $0.1U_{bh}$ in both cases. In the coarser simulations (cf. figure 5.8*a*) the value of the maxima, as well as the value of $(\langle u \rangle - \langle u_p \rangle)/U_{bh}$ in the upper part of the domain is noticeably smaller. This indicates an influence of the resolution on that property in both simulations.

A positive velocity difference $(\langle u \rangle - \langle u_p \rangle)/U_{bh}$ has been observed by several researchers. The experimental evidence of the phenomena covers a range of Stokes numbers, from St_b of order 0.1 in Kiger & Pan (2002) up to Stokes numbers of order 1 in Taniere *et al.* (1997). It was also found to occur in flows with different density ratios (ρ_p/ρ_f), from nearly neutrally buoyant in the case of Kaftori *et al.* (1995) to high density ratios in Taniere *et al.* (1997). Furthermore, it has been observed over a range of different solid volume fraction, from as little as 0.023% in Kaftori *et al.* (1995) up to 2.2% in the present case. The effect of the velocity lag of the solid phase was linked to that the particles are preferentially located in low-speak regions.

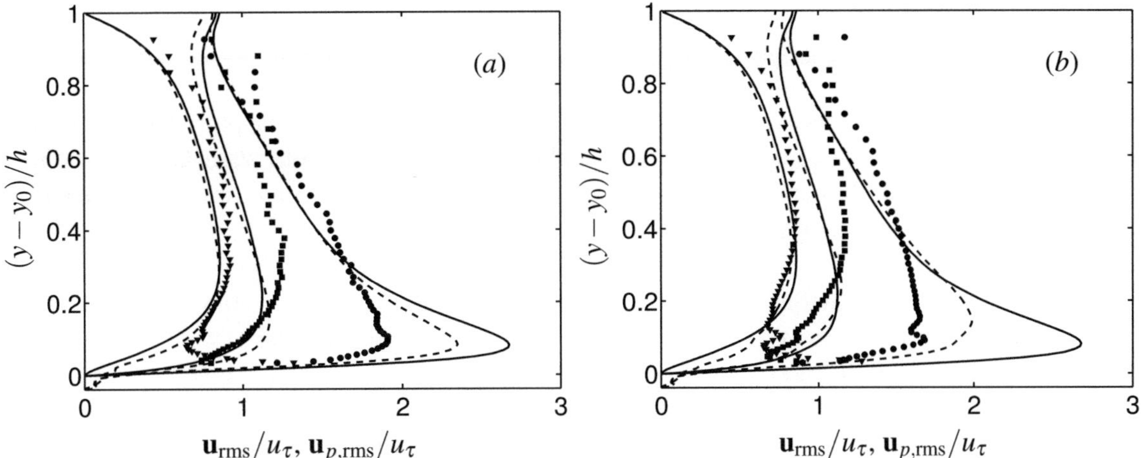

Figure 5.9: Root-mean-square of velocity fluctuations of fluid phase, $\mathbf{u}_{\mathrm{rms}}/u_\tau$ (- - -) and of solid phase, $\mathbf{u}_{p,\mathrm{rms}}/u_\tau$ (symbols) of case T03C (*a*) and case T14C(*b*), in comparison with results of a single-phase open channel flow over a smooth wall (———). Lines and symbols show from left to right wall-normal velocity fluctuation (v_{rms}, $v_{p,\mathrm{rms}}$ ▼), spanwise velocity fluctuation (w_{rms}, $w_{p,\mathrm{rms}}$ ■) and streamwise velocity fluctuation (u_{rms}, $u_{p,\mathrm{rms}}$ •).

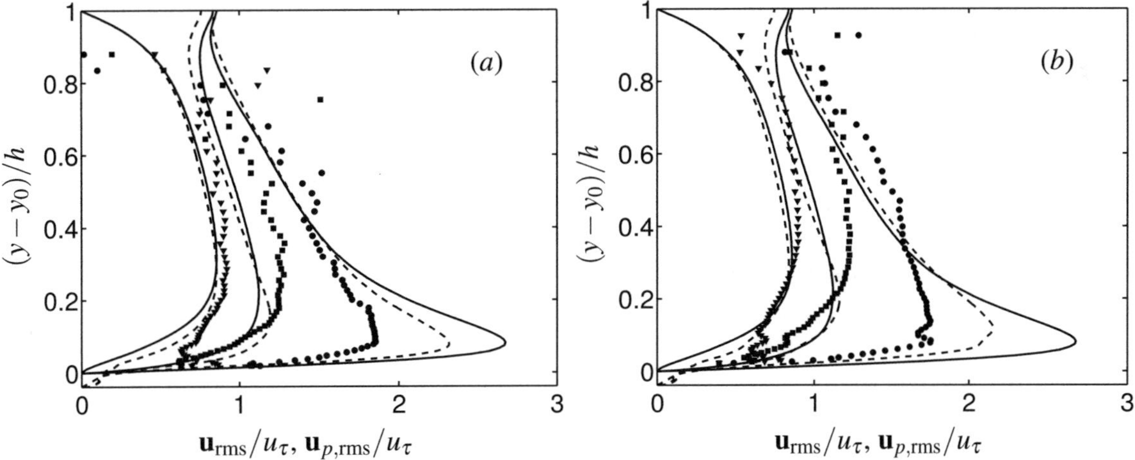

Figure 5.10: Root-mean-square of velocity fluctuations of fluid phase, $\mathbf{u}_{\mathrm{rms}}/u_\tau$ (- - -) and of solid phase, $\mathbf{u}_{p,\mathrm{rms}}/u_\tau$ (symbols) in case T03 (*a*) and case T14 (*b*). in comparison with results of a single-phase open channel flow over a smooth wall (———). Lines and symbols show from left to right wall-normal velocity fluctuation (v_{rms}, $v_{p,\mathrm{rms}}$ ▼), spanwise velocity fluctuation (w_{rms}, $w_{p,\mathrm{rms}}$ ■) and streamwise velocity fluctuation (u_{rms}, $u_{p,\mathrm{rms}}$ •).

The profiles of the velocity fluctuations of both phases in comparison with the velocity fluctuations in the smooth wall simulation are given in figures 5.9 and 5.10. First, focus is given on the velocity fluctuations of the fluid phase. The effect of the moving particles on the velocity statistics of the fluid is similar to the effect of roughness on a single-phase flow: close to the wall the near-wall peak of u_{rms} decreases while the fluctuations v_{rms} and w_{rms} increase somewhat or remain at a similar level compared to the smooth wall simulation. Thus, as in case of roughness, the moving particles lead to a more isotropic turbulence when compared to a single-phase fluid over a smooth wall. Overall, the present results agree well with the experiment of Kiger & Pan (2002) performed in a similar parameter range. One exception it decrease of the near-wall peak value of u_{rms} which is smallest for case T14. While Kiger & Pan (2002) similarly found a decrease it is less pronounced in the experimental results which could be related to the smaller volume fraction used. Above $(y-y_0)/h = 0.4$ the single and two-phase fluid statistics overlap except for the region close to the upper boundary. While the wall-normal component overlaps with smooth wall simulation, in particular the spanwise fluctuations differ from the smooth wall results and are noticeably smaller. This could indicate that the difference between the single-phase flow and the two-phase flow in this region mainly caused by a modification of the large-scale turbulent structures, rather than due to the effects on the small scales.

After the discussion of the fluctuations statistics of the carrier phase, the fluctuations of the disperse phase are discussed in the following. Near the wall, the values of $u_{p,\text{rms}}$ and $w_{p,\text{rms}}$ are smaller than the fluid counterparts, while the value of $v_{p,\text{rms}}$ is found to be larger. The observed increase in $v_{p,\text{rms}}$ compared to v_{rms} is in agreement with the results of Kaftori *et al.* (1995). However, Kiger & Pan (2002) did not observe such an increase. The present results reveal a higher isotropy of the particle fluctuations compared to the fluctuation of the fluid phase. A possible explanation might be the high concentration of particles near the wall. From the value of the solid volume fraction in the present setup, so called four-way coupling can be expected (cf. Elgobashi, 2006). That is, not only particle-turbulence interaction are of importance to the statistics of flow and particles, but also particle-particle and particle-wall contacts. The increased particle collisions close to the wall might lead to the observed increase of isotropy of the particle fluctuations when compared to the fluid fluctuations. Conversely, in the outer region (above $(y-y_0)/h \geq 0.4$) the particle velocity fluctuation components appear to be larger than the fluid counterparts. This might be explained by the increased inertia of the disperse phase and deserves more detailed investigation in future studies.

In figure 5.11 the profiles of the cross-correlation coefficient between streamwise and wall-normal velocity fluctuations (Reynolds stress) are shown for both, fluid and solid phases. The statistics of the fluid phase are found to collapse with the single-phase results, in case T03C and case T03 over the entire flow depth, in case T14C and case T14 for $0.3 \leq (y-y_0)/h$. Thus, for the given range the disperse phase has only little effect on the distribution of the Reynolds stress. In case T14C and case T14 the Reynolds stress of the two-phase flow is found smaller than the smooth wall result for $0.3 \leq (y-y_0)/h$, which might be due to some sort of interference between the particles and the buffer layer structures. The latter might be damped by the presence of the particles and reduce the correlation of the streamwise and wall-normal velocity.

The cross-correlation coefficient of disperse phase, $\langle u'_p v'_p \rangle$, exhibits strong deviations to the fluid Reynolds stress, $\langle u'v' \rangle$. While in the case of lower (higher) global solid volume fraction the particle motion in the streamwise and wall-normal direction is less correlated than the fluid motion for $(y-y_0)/h \leq 0.25$ (0.4), it appears to be higher correlated for $(y-y_0)/h \geq 0.25$ (0.4). These results confirm similar findings in the experiments of Kiger & Pan (2002). Note, that in case T03, the scatter for $(y-y_0)/h \leq 0.25$ is large and the particle velocity correlation is observed higher and lower than the fluid velocity correlations. Due to the scatter it is difficult to infer about its trend with the current data. Different explanations of the result above are possible, in particular the well-known crossing-

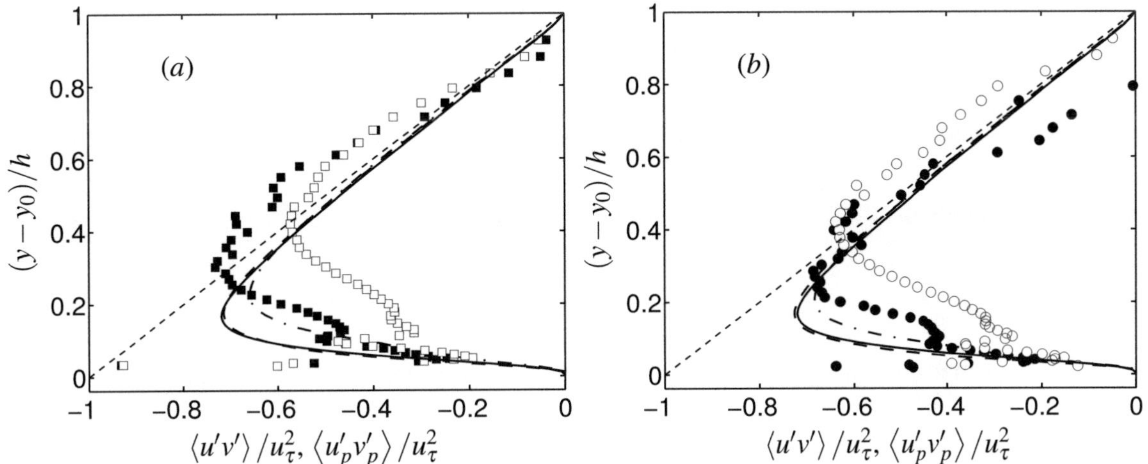

Figure 5.11: Cross-correlation coefficient of streamwise and wall-normal velocities of fluid phase (lines) and particles (symbols) in comparison with a single-phase open channel flow over a smooth wall (———) as a function of wall distance normalised by u_τ^2. (*a*) Case T03C (– – –, ■) and case T14C (– · –, □), (*b*) case T03 (– – –, •) and case T14 (– · –, ○).

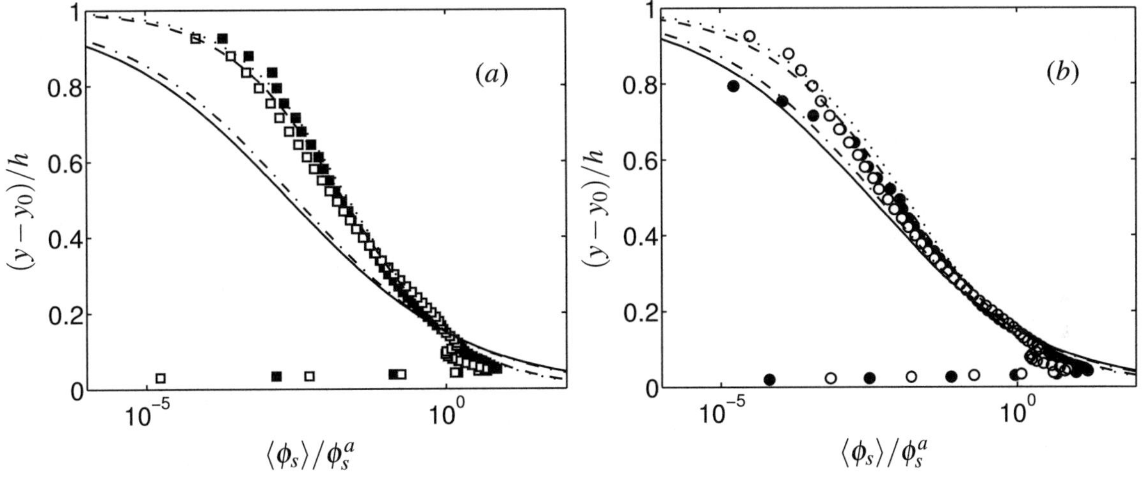

Figure 5.12: Wall-normal profile of particle concentration in comparison with Rouse formula (5.1). (*a*) Case T03C (■) and case T14C (□) in comparison with original definition of β (T03C ———, T14C – · –) and using a 35% lower value of β (T03C – – –, T14C ······). (*b*) Case T03 (•) and case T14 (○) in comparison with original definition of β (T03 ———, T14 – · –) and using a 20% lower value of β (T03 – – –, T14 ······).

trajectories effect of heavy particles in turbulent flow (Yudine, 1959). The observation deserves further consideration in future studies.

5.2.4 Implications for sediment transport

A quantity of high relevance for engineers dealing with equilibrium suspensions is the mean concentration profile of the disperse phase. An expression for the mean concentration can be derived by modelling the distribution of the disperse phase with a convection-diffusion equation (García, 2008,

Case	a/H	ϕ_s^a	β	β_{mod}/β
T03C	0.15	0.008	3.44	0.65
T03	0.15	0.005	3.31	0.80
T14C	0.15	0.053	3.28	0.65
T14	0.15	0.042	3.14	0.80

Table 5.3: Values of the Rouse profiles as defined by (5.1) jointly with a reduced value of β, β_{mod}, as shown in figure 5.12.

pp. 107). Further modelling assumptions in the derivation process include the assumption of a logarithmic law of the wall for the mean streamwise velocity of the fluid phase and an isotropic eddy viscosity model for the turbulent concentration flux. This later assumption is particularly questionable in light of the high particle inertia. The modelling approach leads to the following law for the concentration of the dispersed phase

$$\frac{\langle \phi_s \rangle}{\phi_s^a} = \left(\frac{(h-\tilde{y})/\tilde{y}}{(h-a)/a} \right)^{\beta}, \tag{5.1}$$

where $\tilde{y} = (y-y_0)/h$, the parameter β is defined as $\beta = U_s/(\kappa u_\tau)$, with the 'nominal' settling velocity of the particle, U_s and the near-bed reference concentration, ϕ_s^a, measured at a distance a. The nominal settling velocity of the particle, U_s, is approximated by a balance between drag and immersed weight, using the standard drag formula (Clift *et al.*, 1978), the von Kármán constant, $\kappa \approx 0.41$, and the friction velocity, u_τ, as defined in the simulations. Equation (5.1) is often called the Rousean[1] distribution for suspended sediment.

The concentration profiles of the disperse phase, $\langle \phi_s \rangle / \phi_s^a$, as a function of wall distance, $(y-y_0)/h$, are shown in figure 5.12. The figure additionally contain approximations of the concentration profiles by the Rousean distribution using two different values for the exponent, i.e. the parameter as originally defined β and a modified parameter β_{mod}. The respective parameter used in the approximations employing (5.1) are provided in table 5.3. All approximations employ a value of $a/H = 0.15$ which corresponds to about three particle diameters. Figure 5.12 indicates that the exponent β as defined in (5.1) does not provide a good approximation for the present particle distribution. This might be due to the high degree of simplification inherent to the formula, in particular the modelling of the vertical turbulent concentration flux by means of a simple eddy viscosity. A further evaluation with a modified exponent, β_{mod}, 20% (figure 5.12*a*) and 35% (figure 5.12*b*) lower in value, leads to a satisfactory approximation of the present data for $(y-y_0)/h \geq 0.05$.

Figure 5.4 indicates that particle dispersion is enhanced in the coarse simulations when compared to the fine simulations. This is also partially reflected by the lower ratio of β_{mod}/β in the coarse simulations compared to the fine simulations (cf. table 5.3) and indicates a strong influence of the grid resolution. Also the larger repulsion length of the contact model in the coarse simulations might enhance the particle dispersion. Recall that the definition of the repulsion potential in the particle contact model is $2\Delta_x$ and thus depends on the grid resolution (cf. §2.1.2.2).

[1]Hunter Rouse, American researcher and hydraulic engineer, ⋆29 March 1906 † 16 October 1996

5.3 Summary and recommendation for future work

Interface-resolved direct numerical simulation of sediment transport in horizontal open channel flow at two global solid volume fractions were performed, each with a coarse and a fine resolution of flow field and particles. The considered flow configurations are similar to those of the experiments of Kiger & Pan (2002). The major differences of the present configuration to the latter reference, are that here higher global solid volume fractions are considered jointly with a rough wall consisting of fixed spherical particles in square arrangement, analogously to the simulation of flow over fixed spheres reported in chapter 3.

The presence of particles strongly modifies the mean fluid velocity profiles similar to the effect of roughness on a single phase flow which increases for the present cases with the global solid volume fraction. In agreement with previous experiments in the literature, the mean streamwise velocity of the disperse phase is found to be lower than that of the carrier phase across the entire channel height. In the simulations with sediment transport the streamwise fluid velocity fluctuations are decreased in the buffer layer, while the other fluid velocity fluctuations are enhanced. Once more this effect of moving particle on the fluid flow is similar to the effect of roughness on single phase flow. The trend to more isotropy it even more pronounced for the particle velocity fluctuations which appear to be smaller than the fluid counterpart near the wall and larger in the outer flow. In the simulations with the highest solid volume fraction the fluid Reynolds stress is found to be damped near the wall and essentially unchanged in the outer flow. In the other simulations the Reynolds stress profiles of the two-phase simulations overlaps with the smooth wall simulation over the entire channel height. The cross-correlation of streamwise and wall-normal particle velocity is much lower than the fluid counterpart in the near-wall region, while it appears to be higher outside for the converged simulations. This aspect as well as the trend to higher isotropy might be explained by the increased inertia of the disperse phase but deserves more detailed investigation in future studies.

The concentration profiles were compared with the Rousean distribution which provides a good approximation to the present data when the exponent in the original definition is decreased by a 20% to 35%. The additional modification might reflect the strong simplifications employed during the derivation process of the Rousean distribution. The grid resolution was found to have an influence on some statistics, such as the mean volume fraction as a function of wall-distance, while other quantities, such as the velocity fluctuations, were found to differ little.

To conclude, the present chapter provides highly resolved statistical data of sediment transport in open channel flow. Some potential mechanisms of turbulence-particle interaction have been discussed in the text. Additional statistical analysis is necessary to study these mechanisms in more detail which is a task for future research. In particular it will be important to established whether the details of the numerical treatment of particle collisions affect the results presented here significantly. The different observations suggest that particle inertia, finite-size and finite-Reynolds number effects together with gravity play an important role in the studied flow configurations.

Chapter 6

Summary, conclusions and recommendation for future work

At the moment, I am even more pleased than my wife to have completed the writing.

Albert Alan Townsend

Open channel flow, sediment erosion and sediment transport are of particular interest for hydraulic engineering. For example the control of rivers and hydraulic structures, such as weirs or dams, require formulae to estimate the flow behaviour, the onset of sediment erosion and sediment transport rate. Current formulae are mostly based on empirical relations and on bulk quantities of the flow. The predictive power of these formulae is low. This reflects to some degree a lack of sound understanding of the underlying physics and a need for data with sufficient resolution of flow field and particle related quantities. This thesis contributes to close these gaps of knowledge by providing and analysing high-fidelity data of open channel flow, sediment erosion and sediment transport. The data is generated by direct numerical simulations employing an immersed boundary method to resolve the particle boundary. Three different configurations are studied: (i) open channel flow over fixed spheres, (ii) open channel flow with mobile eroding particles and (iii) open channel flow with sediment transport of many mobile particles. The analysis of the data focused on three aspects: (*a*) the characterisation of force and torque on fixed particles and the related flow structures, (*b*) the characterisation of events that lead to sediment erosion based on results of fixed sphere simulations and simulations of erosion events with mobile particles and (*c*) the effect of many mobile particles on the statistics of flow field and particle related quantities. A brief summary with conclusions on each of these aspects is given below, followed by recommendations for future work. A more detailed discussion on the aspects (*a*–*c*) is provided at the ends of chapters 3 to 5.

6.1 Open channel flow over fixed spheres

Direct numerical simulations of open channel flow over a layer of fixed spheres in square arrangement were performed at a bulk Reynolds number of $\mathrm{Re}_b \approx 2900$. Two cases have been considered. In the first case, the spheres are small (with diameters equivalent to 10.7 wall units) and the limit of the hydraulically smooth flow regime is approached. In the second case, the spheres are more than three times larger (49.3 wall units) and the flow is in the transitionally rough flow regime.

The flow field was studied with respect to its space and time-averaged statistics, the influence of roughness as well as the statistics of its three-dimensional time-average. The analysis of the flow over fixed spheres focused on the characterisation of force and torque on a particle and the related flow structures. The thesis provides a discussion of the statistics of force and torque, the correlation functions of force and torque in time and space-time as well as the correlation functions of force and torque fluctuations to flow field fluctuations. It is found, that the mean values of streamwise force (drag) and wall normal force (lift) are positive, while the mean spanwise torque is negative. The mean values are to a large extent produced at the top region of the particle surface. The intensity of the particle force fluctuations is significantly larger in the large-sphere case, while the trend differs for the fluctuations of the individual components of the torque. For the present simulations, time scales of force and torque fluctuations are of the order of outer flow units. In agreement with this, the correlation of particle force and torque fluctuations and flow field fluctuations have length scales comparable to the channel depth. Flow structures related to force and torque fluctuations appear to travel with a convection velocity of 46% to 71% of the bulk velocity in the small sphere case. Lowest convection velocities were obtained in the large sphere case, which agrees with the finding in the literature that roughness reduces the convection velocity of structures. Some aspects of force and torque on the particles can be explained by simplified considerations that link flow structures to particle force and torque.

6.2 Onset of sediment erosion

The onset of sediment erosion was studied based on the simulations of flow over fixed and mobile eroding spheres. First, several approaches were assessed, to predict the onset of sediment erosion by force and torque on fixed spheres. It was found, that criteria based on high drag, lift or spanwise torque might lead to different results. In contrast to this, the criteria based on drag are, for the present cases, equivalent to more elaborate criteria such as the force tangential to the plane of support as well as the moment around the axis of support. Next, conditionally averaged time-signals of drag, lift and spanwise torque on fixed particles were studied. As conditioning criteria served drag, lift and spanwise torque with a threshold of five standard deviation higher, in case of torque lower, than the mean. It was found, that the shape of the conditionally averaged time-signals heavily depend on the conditioning criteria. In particular, the conditionally averaged time-signals resemble in shape the correlation (in time) between the respective conditioning quantity and the respective averaging quantity. Instantaneous flow fields, conditioned to high drag on fixed particles, often exhibit a pair of high speed and low speed streaks in the vicinity of the particle. Also, strong vortices as identified by highly negative pressure fluctuations are observed near-by the particle and the region around the particle is often characterised by high turbulence activity. The conditionally averaged flow field around particles with high drag show large elongated structures of positive streamwise velocity fluctuations above the particles. The conditionally averaged velocity fluctuations in wall-normal are directed towards the particle while in spanwise direction they are directed away from the particle. As in the case of time signals, the conditionally averaged flow field resembles in shape the correlation between flow field and particle drag. This indicates, that flow structures which might be related to the onset of sediment transport do not differ much from structures that are commonly related to force events.

The potential for sediment erosion of fixed spheres can be assessed by a Shields number based on instantaneous lift. For the present cases, this approach predicts lower critical Shields number for the large sphere case which is in contrast to experimental evidence in the literature. A possible explanation of this discrepancy is that in this approach possible collective effects during erosion are not taken into account. The few direct numerical simulation of erosion events with mobile particles indicate that these collective effects might play a role in the present cases. The numerical simulation of erosion events with mobile particle further show, that during erosion, particles are strongly accelerated in wall-normal direction and along their spanwise axis. Wall normal velocities of the order of the friction velocity are reached within one or two outer flow units. The few studied flow fields during erosion events for the small sphere case exhibit similar flow structures as the flow fields related to high drag events in the fixed sphere case.

6.3 Sediment transport in open channel flow

Direct numerical simulations of particle-laden open channel flow have been performed in which the particle-fluid interface was fully resolved. The numerical setup is similar to the one of flow over small fixed particles. Here however, additionally 2000 and 9216 mobile particles were introduced, leading to a global solid volume fraction of $3.0 \cdot 10^{-3}$ and $1.4 \cdot 10^{-2}$, respectively. Gravity ($|g|\,h/U_{bh}^2 = 0.7$) was applied perpendicularly to the wall and the ratio of particle density to fluid density was 1.7. As a result of gravity, the particles tend to accumulate near the bed. However, in the chosen parameter range, the turbulent motions lead to a cycle of re-suspension and deposition which results in a mean particle concentration profile that decreases with wall-normal distance. It is found, that this profile can be approximated well by the Rousean profile when the exponent in the formula is reduced by 20% to 35%. The presence of particles strongly modifies the mean fluid velocity and the turbulent fluctuation profiles. On average the dispersed phase lags the carrier phase across the whole channel height. Both observations confirm previous experimental evidence. The different observations suggest, that particle inertia, finite-size and finite-Reynolds effects together with gravity play an important role in the studied flow configuration.

6.4 Recommendation of future work

This thesis provides high-fidelity data of two simulations of open channel flow over fixed spheres within the transitionally rough regime. Additional experimental or numerical studies are needed to further explore in particular the fully-rough flow regime and draw conclusions on the effect of the Reynolds number. In future studies, it would be highly beneficial to consider setups in a parameter range that overcomes some problems of existing studies. Most works related to flow over fixed spheres or to the onset of sediment erosion considered small aspect ratios of channel height to particle diameter. Thus, often the flow has to be considered as flow over large obstacles rather than flow over a rough surface. However, a large aspect ratio of channel height to particle diameter is required to establish a certain level of independence of the turbulent flow from the geometry of the wall and thus allow for generalisation of the results. Note that, Jiménez (2004) proposes channel heights of more than 50 characteristic particle sizes. Furthermore, experimental studies are needed that realise large aspect ratios of channel width to channel height. Some evidence exists, that the aspect ratio of 5 proposed by (Nezu & Nakagawa, 1993, p. 97) to assume negligible effects of the wall in the channel centre is to small and higher aspect ratio are needed (cf. figure 4.1 and 4.3 in Detert, 2008 with a ratio of about 4.5, figure 6.24 in Cameron, 2006 with a ratio of about 5). The parameter range for future studies outlined above poses a challenge to both, experimental and numerical studies.

The present study identified a need for data related to force and torque on fixed particles immersed in a layer of spheres. In particular additional experiments at higher Reynolds numbers would be highly beneficial to assess scaling aspects of force and torque and related structures. It might be preferable if direct measurement techniques of force (and torque) would be utilised in such studies. This might be the most accurate method to measure force the more as it was found here, that the distribution of stresses which contribute to force and torque vary strongly across the particles surfaces. Thus measurement techniques that estimate the force on a particle based on single point measurements along the particle surface might be subject to large uncertainties. More insight to scaling aspects based on existing data, could be gained from analysing flow structures related to force and torque on a square surface element in smooth wall flows. Clearly, the force and torque on a particle involves a larger range of complexity and differ in particular with respect to the effect of pressure from the force and torque on a square surface element. However, a smooth wall analogy was found to be very useful in particular for spheres of diameters small with respect to the viscous length. An analogous analysis to the one presented in this thesis, based on the data provided by del Álamo & Jiménez (2003) and Hoyas & Jiménez (2008) in channel flow at moderate Reynolds number would be an inexpensive possibility to assess aspects of Reynolds number effect and scale separation and should be considered in future research. This might also be interesting in the context of resent publications of the experimental studies of the relation of wall shear stress to flow structures based on finite size direct measuring techniques.

With respect to the studies of sediment erosion of mobile particles, additional experimental and numerical studies are needed. A deeper understanding of the processes involved can only be gained by providing enough samples of erosion events to carry out statistical analysis. Experimentally, the approach of Hofland (2005) appears to be an attractive option to reach this goal. The author used a spring to move a particle into its initial position after erosion by rotational motion. Furthermore, the spring was used to trigger the recording of the erosion movement. Numerically, the simulation of the erosion of single spheres as presented in section 4.4 might be seen suited to reduce the numerical costs as several, only weakly correlated, erosion events can be studied within one simulation. However, before considering additional numerical simulations of similar erosion events, future studies should clarify the influence of the contact model on the sediment erosion.

In the future, the simulations of sediment transport presented here should be further analysed to understand some aspects of the flow and particle statistics presented. Questions of interest, which were only addressed briefly here, are the effect of the moving phase on the turbulence structure, possible explanations of the velocity lag of the particle and the higher correlation between streamwise and wall-normal particle velocities compared to fluid velocity. Also a comparison of erosion events of particles at rest in simulation with and without sediment transport would be beneficial to reveal if the erosion events are of similar nature or if they differ.

Appendix A

Preliminary works

Irrtümer haben Ihren Wert,
jedoch nur hier und da:
Nicht jeder der nach Indien fährt,
endeckt Amerika.

Erich Kästner

In the first part of the PhD project the immersed boundary method of Uhlmann (2005*a*) was implemented in the in-house code LEOSCC2, which is successor of the code LESOCC (Breuer & Rodi, 1996) developed in the group of Prof. Wolfgang Rodi at the Institute for Hydromechanics. LESOCC2 solves the Navier–Stokes equations in a curvilinear coordinate system using a block-structured finite volume approach. The code uses a predictor-corrector scheme. In the predictor step the momentum equation is solved using a three-step low-storage Runge-Kutta scheme for the time discretisation. In the corrector step a Poisson equation for the pressure correction is solved, followed by the correction of the pressure and the velocity. A second-order central difference scheme is used for the discretisation in space on a collocated grid arrangement. Coupling of the pressure and velocity fields is enhanced by the momentum interpolation of Rhie & Chow (1983). Further discussion of the numerical scheme of LESOCC2 can be found in Hinterberger (2004) and Fröhlich (2006). The code has been used in several publications of which references can be found in Hinterberger (2004), García-Villalba (2006) and Fröhlich (2006), among others. The implementation of the immersed boundary method in LESOCC2 has been partially documented in Chan-Braun (2012). Preliminary studies of the cases of flow over fixed spheres employing this implementation are reported in Braun (2009), Braun *et al.* (2009*a*) and Braun *et al.* (2009*b*). With the involvement of Markus Uhlmann particle simulations with LESOCC2 were discontinued in favour of using the code SUSPENSE developed by Markus Uhlmann. Due to the design of LESOCC2 as a multi-purpose code on body-fitted grid, LESOCC2 has an overhead leading to an increased requirement of computational resources when compared to the single purpose code SUSPENSE. Also, the interpolation of Rhie & Chow (1983) employed in LESOCC2 reduces the order of the numerical scheme of the code. A similar interpolation is not needed in SUSPENSE as it is based on a staggered variable arrangement leading to a higher accuracy.

Appendix B

Reference cases

In the course of the work on this thesis, reference simulations of open channel flow over a smooth wall were conducted using two different codes. If not stated otherwise, the presented results stem from a simulation carried out with the in-house code LESOCC2 (for a description of the code see appendix A). The corresponding simulation has been named S180. An alternative simulation was carried out employing a pseudo-spectral code which solves the Navier–Stokes equations based on the numerical algorithm of Kim *et al.* (1987), i.e. employing Fourier expansion (with dealiasing) in the periodic streamwise and spanwise directions, and Chebyshev polynomials in wall-normal direction. The code was modified in order to allow for a free-slip boundary condition. The corresponding simulation has been named PS-S180. In the following, the results of both simulations are compared to the simulation of Handler *et al.* (1999) (HSLR-S180). Detail on the numerical setup of the three simulation are given in table B.1. A snapshot of the flow field in simulation PS-S180, which illustrates the computational domain, is provided in figure B.1.

Figure B.2(*a*) shows the mean streamwise velocity profile in semi-logarithmic scale and the root mean square value of the velocity fluctuations of simulations S180 and PS-S180 in comparison with the results of Handler *et al.* (1999). In case S180 statistics were accumulated over about $200H/U_{bH}$ in case PS-S180 statistics were accumulated over about $450H/U_{bH}$. In general the agreement between the flow cases is satisfactory. The maximal differences in the mean streamwise velocity profiles of S180 and PS-S180 to the results of HSLR-S180 are $0.3u_\tau$ ($< 2\% U_{bh}$). The root-mean-square values of the velocity fluctuations shown in figure B.2(*b*) differ less than $0.06u_\tau$ ($< 1\% U_{bh}$) among the simulations. Possible sources of the variations in the results are differences in domain size, observation time, grid resolutions or the different methods used.

Case	Re_b	Re_τ	L_x/H	L_y/H	L_z/H	N_x	N_y	N_z	$\tau_c U_{bH}/H$
S180	2880	183	12	1	3	544	128	272	200
PS-S180	2870	183	12	1	3	340	128	168	450
HSLR-S180	2805	180	4π	1	$3\pi/2$	128	129	128	

Table B.1: Setup parameters of reference smooth wall simulations. Parameters defined as in table 3.1. S180 denotes the reference case employing LESOCC2, PS-S180 denotes the reference case employing a pseudo-spectral method, HSLR-S180 denotes the reference data provided by Handler *et al.* (1999). Note, that in case PS-S180 the dealiased resolution is provided.

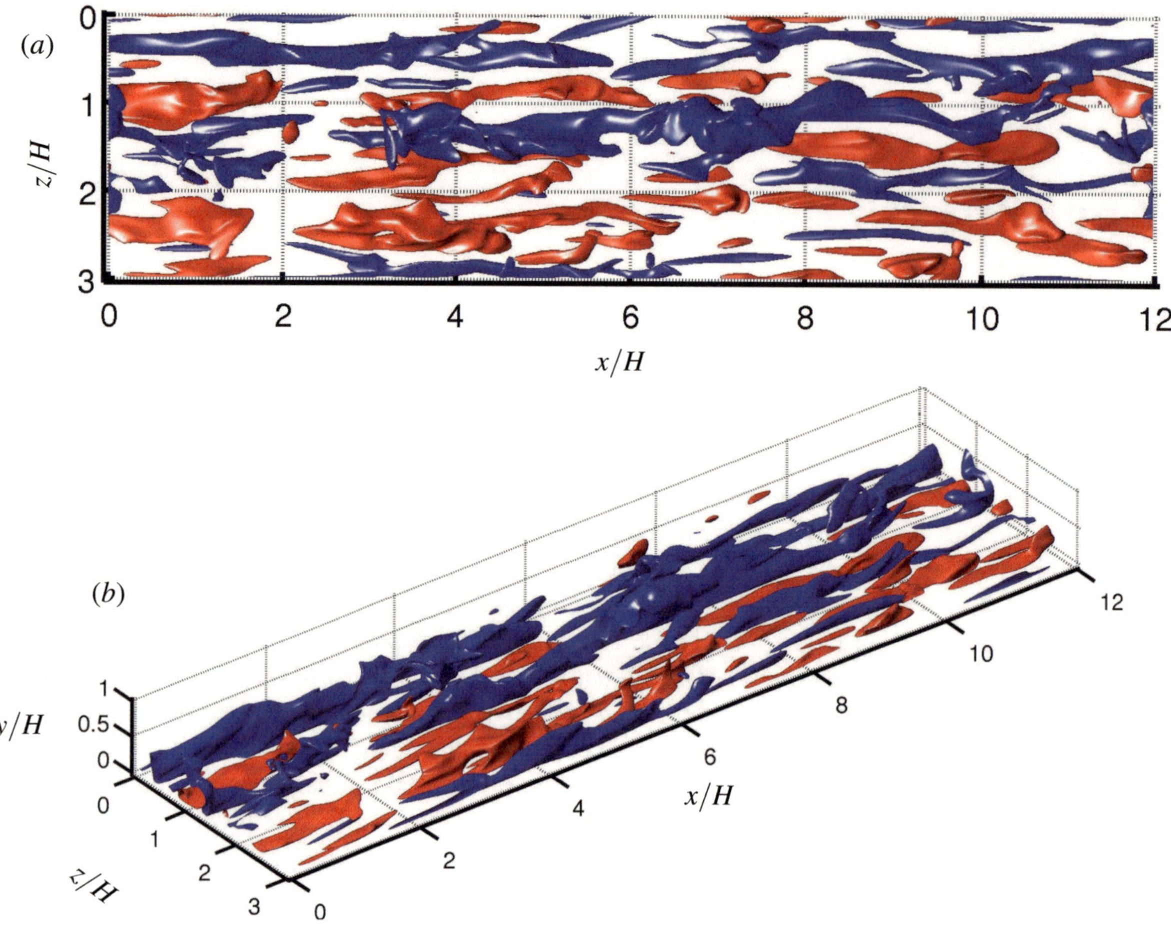

Figure B.1: Instantaneous flow field in smooth wall reference case PS-S180, (*a*) top view (*b*) domain view. Red (blue) surfaces are iso-surfaces of the streamwise velocity fluctuation at values $+3u_\tau$ ($-3u_\tau$).

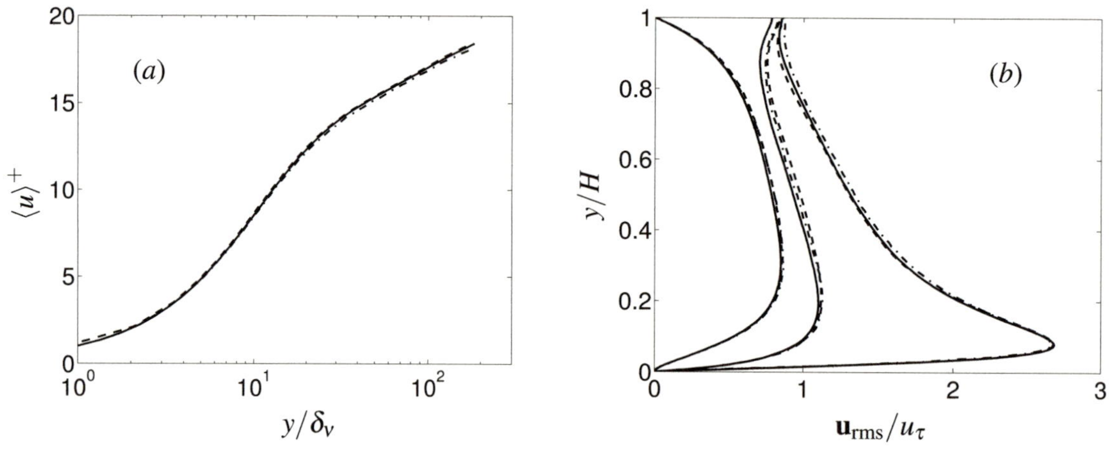

Figure B.2: Results of different smooth wall reference cases, (*a*) mean streamwise velocity profile as a function of wall-normal distance, (*b*) root mean square values of velocity fluctuations as a function of wall-normal distance analogously to figure 3.6(*a*). Lines show S180 (———), PS-S180 (– – –), HSLR-S180 (– · –) described in table B.1.

Appendix C

Extended discussions for fixed spheres cases

C.1 The position of the virtual wall and the friction velocity

Several common methods exist for the definition of an origin y_0 of the wall-normal coordinate when analysing turbulent flow statistics over rough walls. A priori definitions can be based on geometrical considerations. Examples are the volume of the roughness elements divided by the area of the virtual wall (cf. Schlichting, 1936), which for the present geometry leads to $y_0/D = 0.44$ (0.56) in case F10 (F50), or the average of the maximum surface elevation, which leads to $y_0/D = 0.54$ (0.65) in case F10 (F50). A posteriori methods employ the data from measurements or simulations to define y_0. Thom (1971) and Jackson (1981) propose to define y_0 by the wall-normal position of the centroid of the drag profile on the roughness elements. In the present study such a definition would lead to values of $y_0/D = 0.88$ (0.84) in case F10 (F50). It should be noted that in case of a porous sediment layer, this definition is biased by the inter-porous flow. Most researchers, however, use methods which involve the adjustment of a logarithmic law to the mean velocity profile (Raupach *et al.*, 1991), especially for high Reynolds number flows. Based on these methods, several studies on turbulent flow over spherical roughness (for various Reynolds numbers, particle arrangements and flow geometries) can be found that provide the value of y_0 for a given particle diameter (cf. table C.1, reviews in Bayazit, 1983; Nezu & Nakagawa, 1993, p. 26; Dittrich, 1998, p. 29; Detert *et al.*, 2010*a*), including also studies that match well with the present flow conditions (Nakagawa & Nezu, 1977; Grass *et al.*, 1991; Cameron, 2006; Singh *et al.*, 2007). In these studies the virtual wall is positioned at y_0/D in the interval 0.61 to 0.82. In the present work it was chosen to fix the position of the virtual wall at a given level $y_0/D = 0.8$ inside the range of values determined in relevant experiments.

Turning now to the definition of the velocity scale u_τ, three common approaches will be discussed in the following. Again, a widely used method is to obtain u_τ by adjusting a logarithmic law to the mean velocity profile. Assuming the values $\kappa = 0.40$ and $y_0/D = 0.80$ for the von Kármán constant and the offset of the virtual wall, respectively, a fit over the range $50\delta_v \leq (y - y_0) \leq 0.5h$ yields $u_\tau/U_{bh} = 0.062$ (0.081) in case F10 (F50). However, it should be recalled that in the present low-Reynolds number flow the limited extent of the logarithmic region makes this approach relatively error-prone. Alternatively, the global momentum balance can be used in order to relate the driving force (either due to a pressure gradient or gravity) to the different contributions to the drag force generated at the fluid-solid interfaces. While the mean momentum balance is uniquely defined, it does not immediately provide a velocity scale. In some studies the velocity scale is defined from the volumetric force integrated from the virtual wall-distance to the free surface (for example Nakagawa & Nezu, 1977; Detert *et al.*, 2010*a*), i.e. in the present notation $u_\tau^2 = -\langle \mathrm{d}p_l/\mathrm{d}x\rangle h/\rho_f$. This definition leads to $u_\tau/U_{bh} = 0.066$ (0.081) in case F10 (F50).

Finally, let us consider definitions based on the total shear stress profile. In smooth wall flow, the total shear stress τ_{tot} is linear with wall-distance and the appropriate velocity scale is given by $u_\tau^2 = \tau_{tot}(0)/\rho_f$. In rough-wall flow, τ_{tot} in general deviates from a linear relation below the roughness crests which prevents the use of a similar definition, e.g. based upon $\tau_{tot}(y = y_0)$. Instead, some researchers propose to determine the velocity scale independently of the position of the virtual wall by using the total shear stress at the roughness crests, i.e. $u_\tau^2 = \tau_{tot}(y = D)/\rho_f$ (Pokrajac *et al.*, 2006). This definition leads to values of $u_\tau/U_{bh} = 0.066$ (0.080) in case F10 (F50). Note that this latter definition makes a direct comparison of different data sets difficult, since the total shear stress profiles $\tau_{tot}/(\rho_f u_\tau^2)$ represented as a function of $(y-y_0)/h$ will in general not collapse. Alternatively, u_τ can be computed from the total shear stress extrapolated from the region where it varies linearly (i.e. above the roughness crests) down to the position of the virtual wall, yielding the defining relation

$$\tau_{tot} = \rho_f u_\tau^2 \left(1 - \frac{y-y_0}{h}\right), \tag{C.1}$$

valid for $y > D$. This definition leads to $u_\tau/U_{bh} = 0.066$ (0.082) in case F10 (F50).

Incidentally, it can be deduced from the global momentum balance that the present definition implies $u_\tau^2 = -\langle \mathrm{d}p_l/\mathrm{d}x\rangle h/\rho_f$, i.e. it turns out that the definition of u_τ through (C.1) is equivalent to the above mentioned definition used by Nakagawa & Nezu (1977) and Detert *et al.* (2010*a*) based upon an integral of the driving force.

The values obtained for u_τ using the various methods do not differ much. The largest deviations occur for the case F50. It was checked that the results do not change significantly with a different y_0 in the range of values determined in relevant experiments. As an example, figure C.1 shows mean streamwise velocity profile of case F10 and F50 for y_0/D defined in the range of 0.70 to 0.85 in comparison with the smooth wall results. As can be expected, the largest influence can be observed in the vicinity of the rough wall. It can be seen that in both cases the influence on the outer flow is small and does not alter conclusions drawn in this study.

C.2 Details on averaging procedures

In chapter 3 two definitions have been used for averaging a discrete flow field, $\phi(i,j,k)$, in x–z planes

$$\langle\phi\rangle_{xz}^A(j) = \frac{1}{N_x\, N_z}\sum_{i=1}^{N_x}\sum_{k=1}^{N_z}\phi(i,j,k), \tag{C.2}$$

$$\langle\phi\rangle_{xz}^B(j) = \frac{1}{N_m(j)}\sum_{i=1}^{N_x}\sum_{k=1}^{N_z}\phi(i,j,k)\,m(i,j,k), \tag{C.3}$$

where N_x and N_z are the number of grid points in x and z directions, respectively, and m is a field that works as a mask for computing the averages. If a given point lies within the fluid domain, at that point $m = 1$, otherwise $m = 0$. $N_m(j)$ is the sum of h over a wall parallel plane at wall-distance y_j, i.e. $N_m(j) = \sum_{i=1}^{N_x}\sum_{k=1}^{N_z} m(i,j,k)$. $N_m(j)$ equals N_xN_z above the roughness layer such that both expressions (C.3) and (C.2) are equal to each other away from the roughness elements. Within the roughness layer $N_m(j)$ and thus the number of samples for each wall parallel plane decreases. In

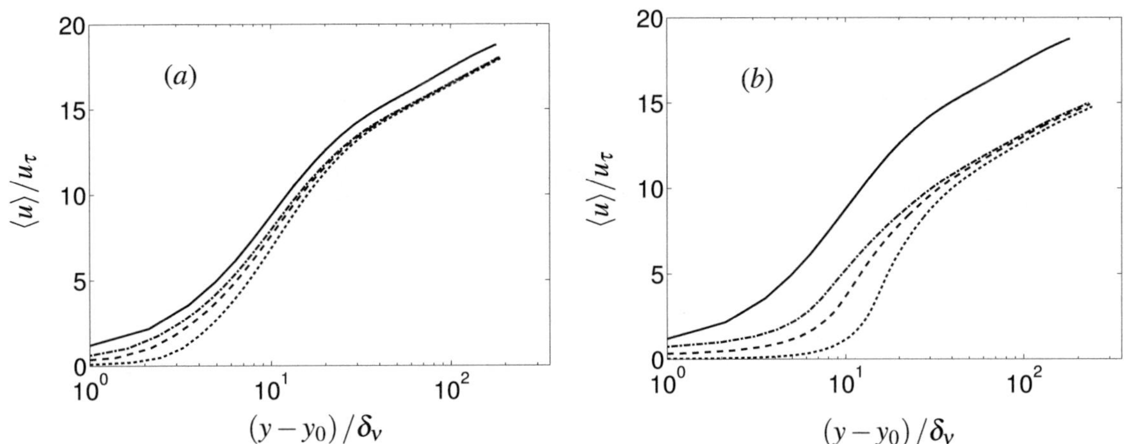

Figure C.1: Time and spatially averaged streamwise velocity component $\langle u \rangle$ of case F10 (*a*) and case F50 (*b*) in comparison with smooth wall open channel flow simulation. Profiles are shown semi-logarithmic scale normalised by δ_ν and u_τ and are the result of different definitions of y_0 and u_τ, $y_0/D = 0.7$ (– · –), $y/D = 0.8$ (– – –), $y_0/D = 0.85$ (······) smooth wall reference case (———).

the context of the double-averaging methodology these two quantities are generally referred to as superficial and intrinsic spatial average, respectively (cf. Nikora *et al.*, 2007).

Note, that the zero-velocity condition is forced only at the surface of the particles due to reasons of efficiency (Uhlmann, 2005*a*). This leads to fictitious non-zero velocities at the grid points that lie within the particles. Fadlun *et al.* (2000) demonstrated that the external flow is essentially unchanged by this procedure which has been confirmed later by Uhlmann (2005*a*). Since the internal fictitious flow affects the value of $\langle \phi \rangle^A_{xz}(j)$ (according to C.2) in the roughness layer, the present averages are computed according to (C.3) where the flow within the roughness layer is discussed (i.e. figures 3.7 and 3.11). When focusing upon the flow above the roughness layer (i.e. in figures 3.4 and 3.6), it was chosen to present data computed according to (C.2), because the number of available samples is larger, as explained in §C.3.

C.3 Consistency of run-time and a posteriori statistics

For the flow field statistics presented in this thesis, two different sets of data have been used. The first set of flow field statistics was collected during the run-time of the simulation employing equation (C.2). This leads to a number of the order of 10^{11} samples per wall-normal grid point in case F10 and F50, collected over the entire observation interval. The second set of data was obtained from analysing stored snapshots of the flow field of which 90 were used in each case. The latter set has been used to compute some additional statistical quantities not stored during run-time. Since it provides a smaller number of samples (roughly a factor of 10^3 less), in the following check its consistency is checked with the more complete set accumulated at run-time.

Figure C.2 shows for each case the second order moments of the velocity fluctuations obtained at run-time in comparison to the same quantities obtained from the snapshots of the simulations applying the averaging operator as defined in (C.2). The differences between the two data sets are small, measuring less than $0.06u_\tau$ ($0.02u_\tau$) in case F10 (F50). Incidentally, it can be seen from the figures

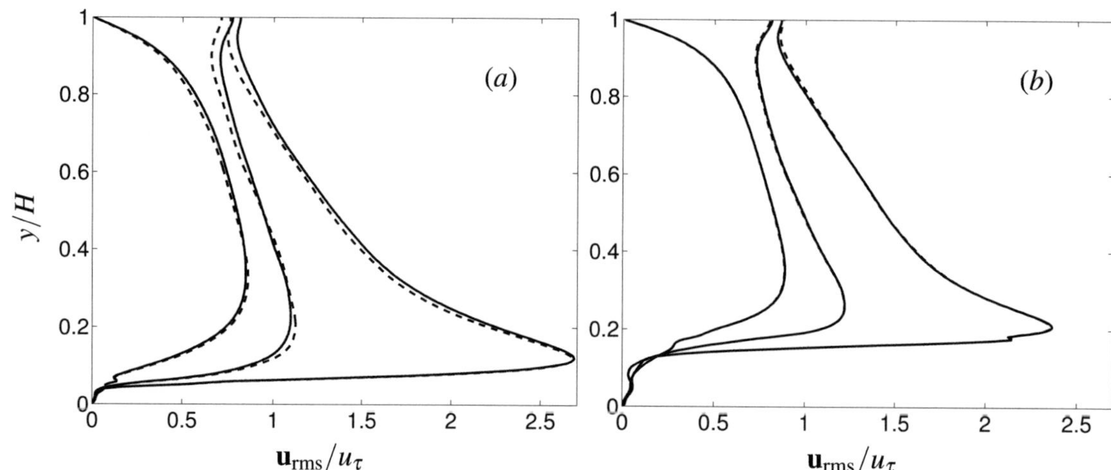

Figure C.2: Comparison of velocity fluctuations normalised by u_τ obtained from run-time (———) and from snapshots (– – –) as a function of y/H in case F10 (*a*) and case F50 (*b*). Curves from left to right are the components in wall-normal (v_{rms}/u_τ), spanwise (w_{rms}/u_τ) and streamwise (u_{rms}/u_τ) direction.

that the discrepancy is largest near the open surface. Therefore it can be concluded, that the data set provided from the 90 stored snapshots is sufficient for the purpose of computing the quantities shown in figure 3.7, figure 3.8(*b*), figures 3.9 to 3.11, and figures 3.38 to 3.46.

Table C.1: Literature on flow over spheres. The literature cover a range of different flow types: closed channel (CC), pipe (P), open channel (OC), boundary layer (BL). Spheres are arranged either in a hexagonal (hex), cubical (cub) or random (rand) packing. The Reynolds number is based on the bulk velocity, the kinematic viscosity and h, which is used here for the closed channel half height, open channel height or the pipe radius respectively. In the boundary layer study the Reynolds number refers to the momentum thickness Reynolds number. D^+ is the particle diameter in viscous scales, W is the channel depth, Δy_0 is the difference between the position of the virtual wall and the particle crests, k_s is the equivalent sand grain roughness.

	type	arrang.	Re [×1000]	D^+	h/D	W/h	$\Delta y_0/D$	k_s/D
Schlichting (1936)	CC	hex	86 – 263	785 – 2100	4.1	10	0.22	0.63
Muñoz Goma & Gelhar (1968)	P	rand	5 – 24	80 – 300	35.8	—	0.23	0.55
Neill (1968)	OC	rand	21 – 32.5	320 – 610	16 – 24	8.5	0.25	—
Nakagawa & Nezu (1977)	OC	rand	8.6 – 10.9	9 – 136	6 – 62	6.2-6.6	0.25	1
Ligrani & Moffat (1986)	BL	hex	3.3 – 18.7	32 – 102	4 – 9	>46	0.18	0.62
Grass *et al.* (1991)	OC	rand	5.1 – 5.6	5 – 107	4 – 43	8.0 – 8.2	0.20 – 0.39	0.68 – 0.85
Cameron *et al.* (2006); Cameron (2006)	OC	hex	5.8 – 10.7	14 – 28	12	3	0.20	—
Manes *et al.* (2007)	OC	cub	5.9 – 35.9	282 – 480	2.3 – 6.5	5 – 14	0	—
Singh *et al.* (2007)	OC	hex	3.11	132	3.2	DNS	0.19	0.77
Detert *et al.* (2010*a*)	OC	hex	20.6	377	5	7.0	0.20	0.81

Appendix D

Definitions of Eulerian statistics of solid phase

Using the sum of a geometric series the boundaries of the bins, $y^b_{\text{bin}}(j)$ are defined as

$$y^b_{\text{bin}}(0) = 0, \quad y^b_{\text{bin}}(j) = \frac{1-(\alpha_{\text{bin}})^j}{1-(\alpha_{\text{bin}})^N_{\text{bin}}} H, \qquad \text{for } j \in [1, N_{\text{bin}}]. \tag{D.1}$$

In the present analysis the number of bins was set to $N_{\text{bin}} = 100$ and the stretching factor is $\alpha_{\text{bin}} = 1.05$. The centres of the bins are defined as

$$y_{\text{bin}}(j) = \left(y^b_{\text{bin}}(j) + y^b_{\text{bin}}(j+1)\right)/2, \qquad \text{for } j \in [1, N_{\text{bin}}]. \tag{D.2}$$

For convenience, $y_{\text{bin}}(j)$ is also denoted y_j or y in the following and throughout the thesis. By means of the bins boundaries, $y^b_{\text{bin}}(j)$, intervals $I_{\text{bin}}(y_j)$, are defined as

$$I_{\text{bin}}(y_j) = \left[y^b_{\text{bin}}(j-1), y^b_{\text{bin}}(j)\right). \tag{D.3}$$

The time and plane averaged number density, n_s, is defined as

$$\langle n_s \rangle (y_j) = \frac{1}{N_t L_x L_z \left(y^b_{\text{bin}}(j) - y^b_{\text{bin}}(j-1)\right)} \sum_{l=1}^{N_t} \sum_{i=1}^{N_p^s} 1, \quad \forall i : y_p^{(i)}(t_l) \in I_{\text{bin}}(y_j), \tag{D.4}$$

where N_t is the number of time steps under consideration, L_x and L_z the periodicity in streamwise and spanwise direction, N_p^s the number of mobile particles i and $y_p^{(i)}(t_l)$ is the wall-normal coordinate of a particle i at time step t_l. The time and plane average of the solid volume fraction, $\phi_s(y_j)$, is defined as

$$\langle \phi_s \rangle (y_j) = V_p \langle n_s \rangle (y_j), \tag{D.5}$$

where V_p is the volume of a particle. The time and plane average of a quantity $\phi_p^{(i)}$ related to a moving particle i is defined as

$$\langle \phi_p \rangle = \frac{1}{\langle n_s \rangle N_t L_x L_z \left(y^b_{\text{bin}}(j) - y^b_{\text{bin}}(j-1)\right)} \sum_{l=1}^{N_t} \sum_{i=1}^{N_p^s} \phi_p^{(i)}(t_l), \quad \forall i : y_p^{(i)}(t_l) \in I_{\text{bin}}(y_j). \tag{D.6}$$

Bibliography

DEL ÁLAMO, J. C. & JIMÉNEZ, J. 2003 Spectra of the very large anisotropic scales in turbulent channels. *Phys. Fluids* **15** (6), L41–L44. 26, 61, 126

DEL ÁLAMO, J. C. & JIMÉNEZ, J. 2009 Estimation of turbulent convection velocities and corrections to Taylor's approximation. *J. Fluid Mech.* **640**, 5–26. 44, 56, 57

DEL ÁLAMO, J. C., JIMÉNEZ, J., ZANDONADE, P. & MOSER, R. D. 2004 Scaling of the energy spectra of turbulent channels. *J. Fluid Mech.* **500**, 135–144. 71

AUTON, T. 1987 The lift force on a spherical body in a rotational flow. *J. Fluid Mech.* **183**, 199–218. 2

AUTON, T. R., HUNT, J. C. R. & PRUD'HOMME, M. 1988 The force on a body in inviscid unsteady non-uniform rotational flow. *J. Fluid Mech.* **197**, 241–257. 2

BAGCHI, P. & BALACHANDAR, S. 2002 Steady planar straining flow past a rigid sphere at moderate Reynolds number. *J. Fluid Mech.* **466**, 365–407. 3

BAGCHI, P. & BALACHANDAR, S. 2003 Effect of turbulence on the drag and lift of a particle. *Phys. Fluids* **15** (11), 3496–3513. 3

BAGCHI, P. & BALACHANDAR, S. 2004 Response of the wake of an isolated particle to an isotropic turbulent flow. *J. Fluid Mech.* **518**, 95–123. 26

BATCHELOR, G. K. 1967 *An introduction to fluid dynamics*, 1st edn. Cambridge: Cambridge University Press. 7, 8

BAYAZIT, M. 1983 Flow structure and sediment transport mechanics in steep channels. In *Mechanics of Sediment Transport, Proc. EUROMECH 156 Colloquium* (ed. B. M. Sumer & A. Müller), pp. 197–206. A. A. Balkema, Rotterdam, Netherlands. 131

BOUSSINESQ, J. 1877 Essai sur la théorie des eaux courantes. *Mémoires présentés par divers savants à l'Académie des Sciences. Paris* **23** (1), 1–660. 13

BRADSHAW, P. 2000 A note on "critical roughness height" and "transitional roughness". *Phys. Fluids* **12** (6), 1611–1614. 21

BRAUN, C. 2009 First results on the impact of turbulent flow on fixed large spherical roughness elements. *Tech. Rep.* 840. Institute for Hydromechanics, Karlsruhe Institute of Technology, University of Karlsruhe, Germany. 127

BRAUN, C., GARCÍA-VILLALBA, M. & UHLMANN, M. 2009*a* A computational study of the hydrodynamics forces on a rough wall. In *Advances in Turbulence XII* (ed. B. Eckhardt), p. 929. 127

BRAUN, C., GARCÍA-VILLALBA, M. & UHLMANN, M. 2009*b* Particle force generation in a turbulent open channel flow. In *33rd IAHR Congress: Water engineering for a sustainable environment*, pp. 44–50. Vancouver, Canada. 127

BREUER, M. & RODI, W. 1996 Large eddy simulation of complex turbulent flows of practical interest. In *Flow simulation with high performance computers II* (ed. E. Hirschel), *Notes on Numerical Fluid Mechanics*, vol. 52, pp. 258–274. Braunschweig: Vieweg. 114, 127

BUFFINGTON, J. M. 1999 The legend of A. F. Shields. *J. Hydraul. Engng* **125** (4), 376–387. 4

BUFFINGTON, J. M. & MONTGOMERY, D. R. 1997 A systematic analysis of eight decades of incipient motion studies, with special reference to gravel-bedded rivers. *Water Resour. Res.* **33** (8), 1993–2029. 4, 101

CAMERON, S. M. 2006 Near-boundary flow structure and particle entrainment. PhD thesis, University of Auckland. 4, 95, 99, 102, 104, 106, 125, 131, 135

CAMERON, S. M., COLEMAN, S. E., MELVILLE, B. W. & NIKORA, V. I. 2006 Marbles in oil, just like a river? In *River Flow 2006* (ed. R. Ferreira, E. Alves, J. Leal & A. Cardoso). Tay. & Fra. Group. 135

CELIK, A. O., DIPLAS, P., DANCEY, C. L. & VALYRAKIS, M. 2010 Impulse and particle dislodgement under turbulent flow conditions. *Phys. Fluids* **22** (4), 046601. 76

CHAN-BRAUN, C. 2012 Implementation of an immersed boundary method in LESOCC2. Report 849. Institute for Hydromechanics, Karlsruhe Institute of Technology. 127

CHAN-BRAUN, C., GARCÍA-VILLALBA, M. & UHLMANN, M. 2010*a* Direct numerical simulation of sediment transport in turbulent open channel flow. In *High performance computing in science and engineering '10* (ed. W. Nagel, D. Kröner & M. Resch). Springer. 109

CHAN-BRAUN, C., GARCÍA-VILLALBA, M. & UHLMANN, M. 2010*b* Numerical simulation of fully resolved particles in rough-wall turbulent open channel flow. In *Proc. 7th Int. Conf. Multiphase Flow* (ed. S. Balachandar & J. S. Curtis). Tampa, USA, CDROM. 109

CHAN-BRAUN, C., GARCÍA-VILLALBA, M. & UHLMANN, M. 2011 Force and torque acting on particles in a transitionally rough open-channel flow. *J. Fluid Mech.* **684**, 441–474. 17

CHAN-BRAUN, C., STREHLE, H., GARCÍA-VILLALBA, M. & UHLMANN, M. 2010*c* Direct numerical simulation of sediment erosion in an open channel flow. In *Gallery of Multiphase Flow, 7th Int. Conf. Multiphase Flow* (ed. S. Balachandar & J. S. Curtis). Tampa, USA, video-clip. 98, 104

CHARRU, F., LARRIEU, E., DUPONT, J.-B. & ZENIT, R. 2007 Motion of a particle near a rough wall in a viscous shear flow. *J. Fluid Mech.* **570**, 431–453. 4

CHERUKAT, P., NA, Y., HANRATTY, T. & MCLAUGHLIN, J. 1998 Direct numerical simulation of a fully developed turbulent flow over a wavy wall. *Theor. Comp. Fluid Dyn.* **11**, 109–134, 10.1007/s001620050083. 2

CHOI, H. & MOIN, P. 1990 On the space-time characteristics of wall-pressure fluctuations. *Phys. Fluids* **2** (8), 1450–1460. 51

CLIFT, R., GRACE, J. R. & WEBER, M. E. 1978 *Bubbles, drops, and particles*. New York [u.a.]: Acad. Pr. 12, 100, 120

COLEBROOK, C. F. 1939 Turbulent flow in pipes with particular reference to the transition region between the smooth- and rouh-pipe laws. *J. Inst. Civil Engrs* **11**, 133–56. 21, 23

CROWE, C. T., SOMMERFELD, M. & TSUJI, Y. 1998 *Multiphase flows with droplets and particles*. Boca Raton, FL: CRC Press. 12

DARCY, H. 1857 *Recherches expérimentales relatives au mouvement de l'eau dans les tuyaux.*. Mallet-Bachelier, Paris. 13

DAVIDSON, P. A. 2007 *Turbulence: an introduction for scientists and engineers*, 1st edn. Oxford: Oxford University Press. 13

DE ANGELIS, V., LOMBARDI, P. & BANERJEE, S. 1997 Direct numerical simulation of turbulent flow over a wavy wall. *Phys. Fluids* **9** (8), 2429–2442. 2

DETERT, M. 2008 Hydrodynamic processes at the water-sediment interface of streambeds. PhD thesis, University of Karlsruhe. 4, 95, 125

DETERT, M., NIKORA, V. & JIRKA, G. H. 2010*a* Synoptic velocity and pressure fields at the water-sediment interface of streambeds. *J. Fluid Mech.* **660**, 55–86. 4, 22, 71, 95, 96, 105, 106, 131, 132, 135

DETERT, M., WEITBRECHT, V. & JIRKA, G. H. 2010*b* Laboratory measurements on turbulent pressure fluctuations in and above gravel beds. *J. Hydraul. Engng* **1**, 126–126. 3

DITTRICH, A. 1998 Wechselwirkung Morphologie/ Strömung naturnaher Fließgewässer. Habilitation, Universität Karlsruhe (TH). 131

DOYCHEV, T. 2010 Grid convergence study of homogeneous particulate flow. Internal Report 846. Institute for Hydromechanics, Karlsruhe Institute of Technology, Karlsruhe. 16

DWIVEDI, A. 2010 Mechanics of sediment entrainment. PhD thesis, The University of Auckland, downloadable at: http://researchspace.auckland.ac.nz. 4, 53, 54, 95

DWIVEDI, A., MELVILLE, B. & SHAMSELDIN, A. Y. 2010 Hydrodynamic forces generated on a spherical sediment particle during entrainment. *J. Hydraul. Engng* **136** (10), 756–769. 4

EINSTEIN, H. A. & EL-SAMNI, E.-S. A. 1949 Hydrodynamic forces on a rough wall. *Rev. Mod. Phys* **21** (3), 520–524. 3, 22, 76

ELGOBASHI, S. 2006 An updated classification map of particle-laden turbulent flows. In *IUTAM Symp. on comp. approaches to multiphase flow* (ed. S. Balachandar & A. Prosperetti), *Fluid mechanics and its applications*, vol. 81, pp. 3–10. Springer Netherlands. 118

FADLUN, E. A., VERZICCO, R., ORLANDI, P. & MOHD-YUSOF, J. 2000 Combined immersed-boundary finite-difference methods for three-dimensional complex flow simulations. *J. Comput. Phys.* **161** (1), 35 – 60. 133

FENTON, J. & ABBOTT, J. 1977 Initial movement of grains on a stream bed: the effect of relative protrusion. *Proc. R. Soc. Lond. A* **352** (1671), 523–537. 102, 106

FLORES, O. & JIMÉNEZ, J. 2006 Effect of wall-boundary disturbances on turbulent channel flows. *J. Fluid Mech.* **566**, 357–376. 2, 18, 25, 26, 58, 64

FLORES, O. & JIMÉNEZ, J. 2010 Hierarchy of minimal flow units in the logarithmic layer. *Phys. Fluids* **22** (7), 071704. 9, 14

FRISCH, U. 1995 *Turbulence: the legacy of A. N. Kolmogorov*, 1st edn. Cambridge: Cambridge University Press. 9, 13

FRÖHLICH, J. 2006 *Large Eddy Simulation turbulenter Strömungen*. Teubner. 127

GARCÍA, M. H. 2008 *Sedimentation engineering: processes, measurements, modeling, and practice*. American Soc. Civil Eng. (ASCE), Reston, Va., ASCE Manual of Practice 110. 4, 32, 97, 119

GARCÍA-VILLALBA, M. 2006 Large-eddy simulation of turbulent swirling jets. PhD thesis, Institute for Hydromechanics, University of Karlsruhe, Karlsruhe. 127

GLOWINSKI, R., PAN, T., HESLA, T. & JOSEPH, D. 1999 A distributed lagrange multiplier fictitious domain method for particulate flows. *Int. J. Multiphase Flow* **25** (5), 755–794. 11, 98

GOLDSTEIN, H., POOLE, C. P. & SAFKO, J. 2002 *Classical mechanics*, 3rd edn. San Francisco: Addison Wesley. 10

MUÑOZ GOMA, R. J. & GELHAR, L. W. 1968 Turbulent pipe flow with rough and porous walls. Int. Rep. 109. Hydrodyn. Lab., Dep. Civil Eng., MIT, Cambridge, Mass. 22, 135

GRASS, A. J., STUART, R. J. & MANSOUR-TEHRANI, M. 1991 Vortical structures and coherent motion in turbulent flow over smooth and rough boundaries. *Philos. Trans. R. Soc. Lond. A* **336**, 36–65. 22, 131, 135

GROSSE, S. & SCHRÖDER, W. 2009 Wall-shear stress patterns of coherent structures in turbulent duct flow. *J. Fluid Mech.* **633**, 147–158. 70

HAGEN, G. 1854 Über den Einfluss der Temperatur auf die Bewegung des Wassers in Röhren. *Abhandlungen der Königlichen Preußischen Akademie der Wissenschaften zu Berlin* pp. 17–98. 13

HALL, D. 1988 Measurements of the mean force on a particle near a boundary in turbulent flow. *J. Fluid Mech.* **187**, 451–466. 3, 32, 35, 151

HANDLER, R. A., SAYLOR, J. R., LEIGHTON, R. I. & ROVELSTAD, A. L. 1999 Transport of a passive scalar at a shear-free boundary in fully developed turbulent open channel flow. *Phys. Fluids* **11** (9), 2607–2625. 129

HINTERBERGER, C. 2004 Dreidimensionale und tiefengemittelte Large-Eddy-Simulation von Flachwasserströmungen. PhD thesis, University of Karlsruhe, Karlsruhe. 114, 127

HOFLAND, B. 2005 Rock and roll, turbulence-induced damage to granular bed protections. PhD thesis, Delft University of Technology. 4, 38, 52, 53, 54, 70, 73, 95, 99, 104, 126

HOFLAND, B. & BATTJES, J. 2006 Probability density functions of instantaneous drag forces and shear stresses on a bed. *J. Hydraul. Engng* **132** (11), 1169–1175. 3

HOFLAND, B., BATTJES, J. & BOOIJ, R. 2005 Measurement of fluctuating pressures on coarse bed material. *J. Hydraul. Engng* **131** (9), 770–781. 3, 28, 52

HOYAS, S. & JIMÉNEZ, J. 2006 Scaling of the velocity fluctuations in turbulent channels up to $\mathrm{Re}_\tau = 2003$. *Phys. Fluids* **18** (1), 011702. 14, 71

HOYAS, S. & JIMÉNEZ, J. 2008 Reynolds number effects on the Reynolds-stress budgets in turbulent channels. *Phys. Fluids* **20** (10), 101511. 61, 126

IKEDA, S. 1982 Incipient motion of sand particles on side slopes. *J. Hydraul. Div.,Proc. ASCE* **108** (1), 95–114. 4

IKEDA, T. & DURBIN, P. A. 2007 Direct simulations of a rough-wall channel flow. *J. Fluid Mech.* **571**, 235–263. 2

JACKSON, P. S. 1981 On the displacement height in the logarithmic velocity profile. *J. Fluid Mech.* **111**, 15–25. 131

JACKSON, R. 2000 *The dynamics of fluidized particles*, 1st edn. Cambridge: Cambridge University Press. 12

JACKSON, R. G. 1976 Sedimentological and fluid-dynamic implications of the turbulent bursting phenomenon in geophysical flows. *J. Fluid Mech.* **77** (03), 531–560. 91

JEON, S., CHOI, H., YOO, J. Y. & MOIN, P. 1999 Space–time characteristics of the wall shear-stress fluctuations in a low-Reynolds-number channel flow. *Phys. Fluids* **11** (10), 3084–3094. 57, 70

JIMÉNEZ, J. 2000 Turbulence. In *Perspectives In Fluid Dynamics* (ed. G. Batchelor, H. Moffatt & M. Worster), pp. 231–288. Cambridge: Cambridge University Press. 13

JIMÉNEZ, J. 2004 Turbulent flow over rough walls. *Annu. Rev. Fluid Mech.* **36**, 173–196. 2, 17, 21, 22, 23, 27, 125, 151

JIMÉNEZ, J. & HOYAS, S. 2008 Turbulent fluctuations above the buffer layer of wall-bounded flows. *J. Fluid Mech.* **611**, 215–236. 40

JIMÉNEZ, J., HOYAS, S., SIMENS, M. P. & MIZUNO, Y. 2010 Turbulent boundary layers and channels at moderate Reynolds numbers. *J. Fluid Mech.* **657**, 335–360. 14

JIMÉNEZ, J. & KAWAHARA, G. 2012 Dynamics of wall-bounded turbulence. *to appear* . 91

JIMÉNEZ, J., KAWAHARA, G., SIMENS, M. P., NAGATA, M. & SHIBA, M. 2005 Characterization of near-wall turbulence in terms of equilibrium and "bursting" solutions. *Phys. Fluids* **17** (1), 015105. 91

JIMÉNEZ, J. & MOIN, P. 1991 The minimal flow unit in near-wall turbulence. *J. Fluid Mech.* **225**, 213–240. 9, 14

JIMÉNEZ, J. & PINELLI, A. 1999 The autonomous cycle of near-wall turbulence. *J. Fluid Mech.* **389**, 335–359. 14

JOHANSSON, A. V., HER, J.-Y. & HARITONIDIS, J. H. 1987 On the generation of high-amplitude wall-pressure peaks in turbulent boundary layers and spots. *J. Fluid Mech.* **175**, 119–142. 105

JOHNSON, K. L. 2003 *Contact mechanics*, 9th edn. Cambridge: Cambridge University Press. 12

KAFTORI, D., HETSRONI, G. & BANERJEE, S. 1995 Particle behavior in the turbulent boundary layer II: Velocity and distribution profiles. *Phys. Fluids* **7** (5), 1107–1121. 5, 116, 118

KENNEDY, J. F. 1995 The Albert Shields story. *J. Hydraul. Engng* **121** (11), 766–772. 4

KIDANEMARIAM, A. G. 2010 Numerical simulation of sediment transport in an open channel flow with fully resolved particles. Master thesis, Institute for Hydromechanics, Karlsruhe Institute of Technology, Karlsruhe. 109

KIGER, K. T. & PAN, C. 2002 Suspension and turbulence modification effects of solid particluates on a hydrizontal turbulent channel flow. *J. Turb.* **3**, 019. 5, 116, 118, 121

KIM, I., ELGHOBASHI, S. & SIRIGNANO, W. A. 1993 Three-dimensional flow over two spheres placed side by side. *J. Fluid Mech.* **246**, 465–488. 3

KIM, J. 1989 On the structure of pressure fluctuations in simulated turbulent channel flow. *J. Fluid Mech.* **205**, 421–451. 24, 50, 51

KIM, J., MOIN, P. & MOSER, R. 1987 Turbulence statistics in fully developed channel flow at low Reynolds number. *J. Fluid Mech.* **177**, 133–166. 14, 24, 25, 26, 40, 64, 129

KING, M. R. & LEIGHTON, D. T. J. 1997 Measurement of the inertial lift on a moving sphere in contact with a plane wall in a shear flow. *Phys. Fluids* **9** (5), 1248–1255. 3

KOLMOGOROV, A. N. 1941 The local structure of turbulence in incompressible viscous fluid for very large Reynolds numbers. *Dokl. Akad. Nauk SSSR* **30**, 9–13, reprinted in *Proc. R. Soc. Lond.* A **434**, 9–13 (1991). 13

KOMORI, S., NAGAOSA, R., MURAKAMI, Y., CHIBA, S., ISHII, K. & KUWAHARA, K. 1993 Direct numerical simulation of three-dimensional open-channel flow with zero-shear gas–liquid interface. *Phys. Fluids* **5** (1), 115–125. 10

KRAVCHENKO, A. G., CHOI, H. & MOIN, P. 1993 On the relation of near-wall streamwise vortices to wall skin friction in turbulent boundary layers. *Phys. Fluids* **5** (12), 3307–3309. 70

KRISHNAN, G. P. & LEIGHTON, D. T. J. 1995 Inertial lift on a moving sphere in contact with a plane wall in a shear flow. *Phys. Fluids* **7** (11), 2538–2545. 3

KROGSTAD, P.-Å. & ANTONIA, R. A. 1994 Structure of turbulent boundary layers on smooth and rough walls. *J. Fluid Mech.* **277**, 1–21. 58

KROGSTAD, P.-Å., KASPERSEN, J. H. & RIMESTAD, S. 1998 Convection velocities in a turbulent boundary layer. *Phys. Fluids* **10** (4), 949–957. 57

LEE, H. & BALACHANDAR, S. 2010 Drag and lift forces on a spherical particle moving on a wall in a shear flow at finite Re. *J. Fluid Mech.* **657**, 89–125. 3, 16

LEONARDI, S., ORLANDI, P. & ANTONIA, R. A. 2007 Properties of d- and k-type roughness in a turbulent channel flow. *Phys. Fluids* **19** (12), 125101. 2

LEONARDI, S., ORLANDI, P., SMALLEY, R. J., DJENIDI, L. & ANTONIA, R. A. 2003 Direct numerical simulations of turbulent channel flow with transverse square bars on one wall. *J. Fluid Mech.* **491**, 229–238. 2

LIGRANI, P. M. & MOFFAT, R. J. 1986 Structure of transitionally rough and fully rough turbulent boundary layers. *J. Fluid Mech.* **162**, 69–98. 21, 23, 135

LOBKOVSKY, A. E., ORPE, A. V., MOLLOY, R., KUDROLLI, A. & ROTHMAN, D. H. 2008 Erosion of a granular bed driven by laminar fluid flow. *J. Fluid Mech.* **605**, 47–58. 4

LOTH, E. expected publication fall 2011 *Computational Fluid Dynamics of Bubbles, Drops and Particles*. Cambridge University Press. 16

LUCCI, F., FERRANTE, A. & ELGHOBASHI, S. 2010 Modulation of isotropic turbulence by particles of Taylor length-scale size. *J. Fluid Mech.* **650**, 5–55. 16

LUCCI, F., FERRANTE, A. & ELGHOBASHI, S. 2011 Is Stokes number an appropriate indicator for turbulence modulation by particles of Taylor-length-scale size? *Phys. Fluids* **23** (2), 025101. 16

MANES, C., POKRAJAC, D. & MCEWAN, I. 2007 Double-averaged open-channel flows with small relative submergence. *J. Hydraul. Engng* **133** (8), 896–904. 135

MARUSIC, I., MCKEON, B. J., MONKEWITZ, P. A., NAGIB, H. M., SMITS, A. J. & SREENIVASAN, K. R. 2010 Wall-bounded turbulent flows at high Reynolds numbers: Recent advances and key issues. *Phys. Fluids* **22** (6), 065103. 2, 71

MIYAKE, Y., TSUJIMOTO, K. & NAKAJI, M. 2001 Direct numerical simulation of rough-wall heat transfer in a turbulent channel flow. *Int. J. Heat Fluid Fl.* **22** (3), 237 – 244. 2

MOIN, P. & MAHESH, K. 1998 Direct numerical simulation: A tool in turbulence research. *Annu. Rev. Fluid Mech.* **30** (1), 539–578. 14, 15

MOLLINGER, A. & NIEUWSTADT, F. 1996 Measurement of the lift force on a particle fixed to the wall in the viscous sublayer of a fully developed turbulent boundary layer. *J. Fluid Mech.* **316**, 285–306. 3, 34

MUSTE, M., YU, K., FUJITA, I. & ETTEMA, R. 2009 Two-phase flow insights into open-channel flows with suspended particles of different densities. *Environ. Fluid Mech.* **9**, 161–186, 10.1007/s10652-008-9102-7. 5

MUTHANNA, C., NIEUWSTADT, F. T. M. & HUNT, J. C. R. 2005 Measurement of the aerodynamic forces on a small particle attached to a wall. *Exp. Fluids* **39**, 455–463. 3

NADEN, P. 1987 An erosion criterion for gravel-bed rivers. *Earth Surf. Process. Landf.* **12** (1), 83–93. 4

NAGANO, Y., HATTORI, H. & HOURA, T. 2004 DNS of velocity and thermal fields in turbulent channel flow with transverse-rib roughness. *Int. J. Heat Fluid Fl.* **25** (3), 393 – 403. 2

NAKAGAWA, H. & NEZU, I. 1977 Prediction of the contributions to the Reynolds stress from bursting events in open-channel flows. *J. Fluid Mech.* **80** (01), 99–128. 22, 131, 132, 135

NEILL, C. 1968 A re-examination of the beginning of movement for coarse granular bed materials. Internal report. Hydraulic Research Station, Wallingfort, UK. 135

NEZU, I. & NAKAGAWA, H. 1993 *Turbulence in Open-Channel Flows*. IAHR/AIRH Monograph Series, Balkema Publishers. 125, 131

NIKORA, V., MCEWAN, I., MCLEAN, S., COLEMAN, S., POKRAJAC, D. & WALTERS, R. 2007 Double-averaging concept for rough-bed open-channel and overland flows: Theoretical background. *J. Hydraul. Engng* **133** (8), 873–883. 133

NIKORA, V. I., GORING, D. G., MACEWAN, I. & GRIFFITHS, G. 2001 Spatially averaged open-channel flow over rough bed. *J. Hydraul. Engng* **127** (2), 123–133. 27

NIKURADSE, J. 1933 Strömungsgesetze in rauhen Rohren. *VDI-Forschungsheft* **361**, engl. translation 1950, Laws of flow in rough pipes. NACA TM 1292. 21, 23

NIÑO, Y. & GARCÍA, M. H. 1996 Experiments on particle–turbulence interactions in the near–wall region of an open channel flow: implications for sediment transport. *J. Fluid Mech.* **326**, 285–319. 4

NIÑO, Y., LOPEZ, F. & GARCÍA, M. 2003 Threshold for particle entrainment into suspension. *Sedimentology* **50** (2), 247–263. 4

O'NEILL, P. L., NICOLAIDES, D., HONNERY, D. & SORIA, J. 2004 Autocorrelation functions and the determination of integral length with reference to experimental and numerical data. In *Proc. 15th Australasian Fluid Mech. Conference*. The University of Sydney, Sydney, Australia. 42

OPPENHEIM, A. & SCHAFER, R. 1989 *Discrete-time signal processing*. Prentice-Hall. 45

ORLANDI, P. & LEONARDI, S. 2008 Direct numerical simulation of three-dimensional turbulent rough channels: parameterization and flow physics. *J. Fluid Mech.* **606**, 399–415. 2, 22

ORLANDI, P., LEONARDI, S., TUZI, R. & ANTONIA, R. A. 2003 Direct numerical simulation of turbulent channel flow with wall velocity disturbances. *Phys. Fluids* **15** (12), 3587–3601. 2

OURIEMI, M., AUSSILLOUS, P., MEDALE, M., PEYSSON, Y. & ÉLISABETH GUAZZELLI 2007 Determination of the critical Shields number for particle erosion in laminar flow. *Phys. Fluids* **19** (6), 061706. 4

PAN, T.-W., JOSEPH, D. D., BAI, R., GLOWINSKI, R. & SARIN, V. 2002 Fluidization of 1204 spheres: simulation and experiment. *J. Fluid Mech.* **451**, 169–191. 11

PAN, Y. & BANERJEE, S. 1997 Numerical investigation of the effects of large particles on wall-turbulence. *Phys. Fluids* **9** (12), 3786–3807. 5

PAPANICOLAOU, A., DIPLAS, P., EVAGGELOPOULOS, N. & FOTOPOULOS, S. 2002 Stochastic incipient motion criterion for spheres under various packing conditions. *J. Hydraul. Engng* **128** (4), 369–380. 4, 32

PESKIN, C. S. 1972 Flow patterns around heart valves: A digital computer method for solving the equation of motion. PhD thesis, Albert Einstein College of Medicine. 15

PESKIN, C. S. 2002 The immersed boundary method. *Acta Numerica* **11**, 479–517. 15

PEYSSON, Y., OURIEMI, M., MEDALE, M., AUSSILLOUS, P. & ÉLISABETH GUAZZELLI 2009 Threshold for sediment erosion in pipe flow. *Int. J. Multiphase Flow* **35** (6), 597 – 600. 4

PIMENTA, M. M., MOFFAT, R. J. & KAYS, W. M. 1975 The turbulent boundary layer: an experimental study of the transport of momentum and heat with the effect of roughness. Internal report HMT-21. Dep. Mech. Eng. Stanford Univ., Stanford, California 94305. 21

POGGY, D., PORPORATO, A. & RIDOLFI, L. 2003 Analysis of the small-scale structure of turbulence on smooth and rough walls. *Phys. Fluids* **15** (1), 35–46. 22

POKRAJAC, D., FINNIGAN, J., MANES, C., MCEWAN, I. & NIKORA, V. 2006 On the definition of shear velocity in rough bed open-channel flows. In *River Flow 2006* (ed. R. Ferreiara, E. Alves, J. Leal & A. Cardoso). A.A. Balkema, Rotterdam. 132

POKRAJAC, D. & MANES, C. 2009 Velocity measurements of a free-surface turbulent flow penetrating a porous medium composed of uniform-size spheres. *Transport Porous Med.* **78**, 367 – 383. 27

POPE, S. 2000 *Turbulent Flows*. Cambridge University Press. 7, 8, 13, 14, 20, 42, 45

PROSPERETTI, A. & TRYGGVASON, G., ed. 2007 *Computational methods for multiphase flow*. Cambridge: Cambridge University Press. 12

QUADRIO, M. & LUCHINI, P. 2003 Integral space–time scales in turbulent wall flows. *Phys. Fluids* **15** (8), 2219–2227. 50, 51, 57

RAO, K. N., NARASIMHA, R. & NARAYANAN, M. A. B. 1971 The bursting phenomenon in a turbulent boundary layer. *J. Fluid Mech.* **48** (02), 339–352. 91

RAUPACH, M. R., ANTONIA, R. A. & RAJAGOPALAN, S. 1991 Rough-wall turbulent boundary layers. *App. Mech. Rev.* **44** (1), 1–25. 18, 131

REYNOLDS, O. 1883 An experimental investigation of the circumstances which determine whether the motion of water shall be direct or sinuous, and of the law of resistance in parallel channels. *Philos. Trans. R. Soc. Lond. A* **174**, 935–982. 13

REYNOLDS, O. 1895 On the dynamical theory of incompressible viscous fluids and the determination of the criterion. *Philos. Trans. R. Soc. Lond. A* **186**, 123–164. 13

RHIE, C. & CHOW, W. 1983 Numerical study of the turbulent-flow past an airfoil with trailing edge separation. *AIAA J.* **21** (11), 1525–1532. 127

RICHARDSON, L. F. 1922 *Weather prediction by numerical process*. Cambridge: Cambridge University Press. 13

RIGHETTI, M. & ROMANO, G. 2004*a* Particle-fluid interactions in a plane near-wall turbulent flow. *J. Fluid Mech.* **505**, 93–121. 5

RIGHETTI, M. & ROMANO, G. P. 2004*b* Particle fluid interactions in a plane near-wall turbulent flow. *J. Fluid Mech.* **505**, 93–121. 5

VAN RIJN, L. C. 1993 *Principles of sediment transport in rivers, estuaries, and costal seas*. Aqua publications, Amsterdam, Netherlands. 4, 97, 99

ROBINSON, S. K. 1991 Coherent motions in the turbulent boundary layer. *Annu. Rev. Fluid Mech.* **23**, 601–638. 91

ROMA, A., PESKIN, C. & BERGER, M. 1999 An adaptive version of the immersed boundary method. *J. Comput. Phys.* **153** (2), 509–534. 15

ROUSE, H. & INCE, S. 1957 *History of hydraulics*. Iowa Institute of Hydraulic Research. 13

SAFFMAN, P. G. 1965 The lift on a small sphere in a slow shear flow. *J. Fluid Mech.* **22** (2), 385–400. 2

SCHLATTER, P. & ÖRLÜ, R. 2010 Assessment of direct numerical simulation data of turbulent boundary layers. *J. Fluid Mech.* **659**, 116–126. 14

SCHLICHTING, H. 1936 Experimentelle Untersuchungen zum Rauhigkeitsprobem. *Ing. Arch.* **7**, 1–34. 21, 131, 135

SCHLICHTING, H. 1965 *Grenzschicht-Theorie*, 5th edn. Karlsruhe: Verlag G. Braun. 2, 8, 20

SHENG, J., MALKIEL, E. & KATZ, J. 2009 Buffer layer structures associated with extreme wall stress events in a smooth wall turbulent boundary layer. *J. Fluid Mech.* **633**, 17–60. 70, 91

SHIELDS, A. 1936 Anwendung der Ähnlichkeitsmechanik und der Turbulenzforschung auf die Geschiebebewegung. *Mitteilungen der Versuchsanstalt für Wasserbau und Schiffbau (Berlin)* **26**. 4, 101

SHOCKLING, M. A., ALLEN, J. J. & SMITS, A. J. 2006 Roughness effects in turbulent pipe flow. *J. Fluid Mech.* **564**, 267–285. 21, 23

SINGH, K. M., SANDHAM, N. D. & WILLIAMS, J. J. R. 2007 Numerical simulation of flow over a rough bed. *J. Hydraul. Engng* **133** (4), 386–398. 2, 22, 131, 135

SMITH, C. R. & METZLER, S. P. 1983 The characteristics of low-speed streaks in the near-wall region of a turbulent boundary layer. *J. Fluid Mech.* **129**, 27–54. 64

SPALART, P. R. 1988 Direct simulation of a turbulent boundary layer up to $Re_\theta = 1410$. *J. Fluid Mech.* **187**, 61–98. 14

STREHLE, H. 2011 Numerical simulation of sediment erosion. Studientarbeit, Institute for Hydromechanics, Karlsruhe Institute of Technology, Karlsruhe. 98, 99, 104

SUTHERLAND, A. J. 1967 Proposed mechanism for sediment entrainment by turbulent flows. *J. Geophys. Res.* **72** (24), 6183–6194. 4

TANIERE, A., OESTERLE, B. & MONNIER, J. 1997 On the behaviour of solid particles in a horizontal boundary layer with turbulence and saltation effects. *Exp. Fluids* **23** (6), 463–471. 5, 116

TAYLOR, G. I. 1938 The spectrum of turbulence. *Proc. R. Soc. London* **164** (919), 476–490. 54

TENNEKES, H. & LUMLEY, J. L. 1972 *A First course in turbulence*. Cambridge (Mass.): MIT Pr. 13

THOM, A. S. 1971 Momentum absorption by vegetation. *Q.J.R. Meteorol. Soc.* **97** (414), 414–428. 131

TOWNSEND, A. A. 1971 *The structure of turbulent shear flow*. Cambridge: Cambridge University Press. 18

TRITTON, D. J. 1988 *Physical fluid dynamics*, 2nd edn. Oxford: Clarendon Pr. 42

UHLMANN, M. 2003*a* First experiments with the simulation of particulate flows. Technical Report No. 1020, CIEMAT, Madrid, Spain, ISSN 1135-9420. 15

UHLMANN, M. 2003*b* Simulation of particulate flows on multi-processor machines with distributed memory. Technical Report No. 1039, CIEMAT, Madrid, Spain, ISSN 1135-9420. 15

UHLMANN, M. 2004 New results on the simulation of particulate flows. *Tech. Rep.* 1038. CIEMAT, Madrid, Spain, ISSN 1135-9420. 15, 16

UHLMANN, M. 2005*a* An immersed boundary method with direct forcing for the simulation of particulate flows. *J. Comput. Phys.* **209** (2), 448–476. 15, 16, 127, 133

UHLMANN, M. 2005*b* An improved fluid-solid coupling method for DNS of particulate flow on a fixed mesh. In *Proc. 11th Workshop Two-Phase Flow Predictions, Merseburg, Germany* (ed. M. Sommerfeld). Universität Halle, ISBN 3-86010-767-4. 16

UHLMANN, M. 2006*a* Direct numerical simulation of sediment transport in a horizontal channel. *Tech. Rep.*. CIEMAT, Madrid, Spain, ISSN 1135-9420. 98, 101

UHLMANN, M. 2006*b* Experience with DNS of particulate flow using a variant of the immersed boundary method. In *Proc. ECCOMAS CFD 2006* (ed. P. Wesseling, E. Oñate & J. Périaux). Egmond aan Zee, The Netherlands: TU Delft, ISBN 90-9020970-0. 15, 16

UHLMANN, M. 2008 Interface-resolved direct numerical simulation of vertical particulate channel flow in the turbulent regime. *Phys. Fluids* **20** (5), 053305. 11, 15, 16, 100, 112

UHLMANN, M. & FRÖHLICH, J. 2007 Transport of heavy spherical particles in horizontal channel flow. In *High performance computing in science and engineering 2007, HLRS, Springer*. 104

VANONI, V. 1975 *Sedimentation engineering*. ASCE–Manuals and reports on engineering practice–No. 54, New York. 4, 99

VERZICCO, R. & ORLANDI, P. 1996 A finite-difference scheme for three-dimensional incompressible flows in cylindrical coordinates. *J. Comput. Phys.* **123**, 402–414. 15

WELCH, P. 1967 The use of fast Fourier transform for the estimation of power spectra: a method based on time averaging over short, modified periodograms. *IEEE Trans. Audio Electroacoustics* **AU-15**, 70–73. 45, 57

WILLETTS, B. B. & MURRAY, C. G. 1981 Lift exerted on stationary spheres in turbulent flow. *J. Fluid Mech.* **105**, 487–505. 3

YALIN, M. 1977 *Mechanics of sediment transport*, 2nd edn. Pergamon Press. 4

YANG, D. & SHEN, L. 2010 Direct-simulation-based study of turbulent flow over various waving boundaries. *J. Fluid Mech.* **650**, 131–180. 2

YUDINE, M. 1959 Physical consideration on heavy-particle diffusion. *Adv. Geophys.* **6**, 185–191. 119

YUN, G., KIM, D. & CHOI, H. 2006 Vortical structures behind a sphere at subcritical Reynolds numbers. *Phys. Fluids* **18** (1), 015102. 3

ZENG, L., BALACHANDAR, S., FISCHER, P. & NAJJAR, F. 2008 Interactions of a stationary finite-sized particle with wall turbulence. *J. Fluid Mech.* **594**, 271–305. 3, 26

ZENG, L., NAJJAR, F., BALACHANDAR, S. & FISCHER, P. 2009 Forces on a finite-sized particle located close to a wall in a linear shear flow. *Phys. Fluids* **21** (3), 033302. 3

List of Tables

List of Figures